EURO Advanced Tutorials on Operational Research

Series Editor

M. Grazia Speranza, Brescia, Italy

The EURO Advanced Tutorials on Operational Research are a series of short books devoted to an advanced topic—a topic that is not treated in depth in available textbooks. The series covers comprehensively all aspects of Operations Research. The scope of a Tutorial is to provide an understanding of an advanced topic to young researchers, such as Ph.D. students or Post-docs, but also to senior researchers and practitioners. Tutorials may be used as textbooks in graduate courses. All titles in the series are peer-reviewed.

This series has been accepted by Scopus.

Celso C. Ribeiro • Sebastián Urrutia •
Dominique de Werra

Combinatorial Models for Scheduling Sports Tournaments

Springer

Celso C. Ribeiro
Universidade Federal Fluminense
Niterói, Rio de Janeiro, Brazil

Sebastián Urrutia
Molde University College
Molde, Norway

Dominique de Werra
École Polytechnique Fédérale de Lausanne
Lausanne, Vaud, Switzerland

ISSN 2364-687X ISSN 2364-6888 (electronic)
EURO Advanced Tutorials on Operational Research
ISBN 978-3-031-37282-7 ISBN 978-3-031-37283-4 (eBook)
https://doi.org/10.1007/978-3-031-37283-4

This Springer imprint is published by the registered company Springer Nature Switzerland AG
The registered company address is: Gewerbestrasse 11, 6330 Cham, Switzerland

Paper in this product is recyclable.

Foreword

Everyday, countless individuals and teams play sports of every kind. And whether it is the final of the FIFA World Cup or a bunch of six-year-olds playing tee-ball, someone has to figure out the who, when, and where of the matches. It is only recently that the theory and practice of sports scheduling has moved to a point where it can have a significant impact on real sports leagues.

I learned of sports scheduling back in 1995 when I was approached to schedule the games for Major League Baseball. With 30 teams and 2430 games, creating a playable schedule is a formidable task. Fortunately, there was some literature on sports scheduling, most notably by one of the authors of this monograph, Dominique de Werra. De Werra practically created sports scheduling as a field of study, examining the fundamental issues of round-robin scheduling and the use of graph theory in this application.

Unfortunately, it became quickly clear that there is no such thing as "the sports scheduling problem." Every league has a different set of requirements and wishes, and no one technique can account for all of them. For Major League Baseball, one major wish was to minimize the amount the teams travel, since teams can travel up to 50,000 miles in a single season. No work had been done on this issue and standard methods, like straightforward integer programming, failed miserably on this problem.

But it turns out that techniques could be developed that worked on different types of leagues. Together with my colleague George Nemhauser, we began scheduling college basketball conferences. Computing power, together with new methods, meant that finally many real leagues could have optimal or near-optimal schedules.

Leading the way for the past 20 years are the other authors of this monograph: Celso Ribeiro and Sebastián Urrutia. Together with other researchers, they have studied a wide variety of sports scheduling problems, both those that arise in practice and those as part of the theoretical underpinning of the field. From soccer and referee scheduling, to understanding the role breaks (consecutive away games) and fixed venues play, to developing exact and heuristic approaches for various sports scheduling problems, Ribeiro and Urrutia's works (together and separately, with numerous other co-authors) have truly led the tremendous growth in sports scheduling research and application.

Up until 1995, there were 18 papers in the literature with the phrase "sports scheduling" in the title (about half by de Werra); since 1995 there are hundreds, with countless more on scheduling particular leagues or particular new problems such as the Traveling Tournament Problem. It is truly time for a monograph to bring together many of the results, and Ribeiro, Urrutia, and de Werra have now provided that monograph.

So what happened with Major League Baseball? It took ten years before the models, techniques, and computing power were sufficient to create playable schedules, and I (along with Nemhauser and Kelly Easton) used those methods in the 2005–2017 period to provide Major League Baseball, its teams, and its umpires with their schedules.

This monograph will be useful to many audiences: if you want to schedule a real league, you may well find the answers in this book; if you are looking for interesting research directions, this book abounds with them; if you want cases on how integer and combinatorial optimization can be used to model and solve problems of practical interest, sports scheduling provides many wonderful examples.

If I had this monograph back in 1995, I could have saved a lot of time!

Pittsburgh, PA, USA
December 2022

Michael Trick

Preface

The project of this book emerged from the previous work of the three authors in the area of sports scheduling. It has matured for some time. The authors have been collaborating since 2015.

Since the beginning, our goal has been to combine fundamental results on tournament scheduling, exact and approximate solution algorithms, and some of their applications in the same manuscript. Some joint journal papers and research visits have been the occasion to put the hands and brains to work and start writing.

And then a pandemic arrived.

The book was delayed for two years.

These were difficult times for our families, friends, colleagues, and students.

For everyone.

We revere the memory of those who suffered and lost their lives. We are grateful to those who worked hard to get us through this period.

We are pleased to see this book appear at a time when, thanks to Science, the World begins to find a new paradigm of normality.

Although some highly cited surveys about sports scheduling exist, this is the first self-contained textbook about this subject. It describes fundamental combinatorial structures that represent round-robin tournament scheduling problems, exact and approximate algorithms for their solution, and real-life applications.

This book is aimed at students, engineers, computer scientists, operations researchers, sports managers, and other professionals interested in sports scheduling fundamentals, algorithms, and applications. Parts of this book have been presented in talks and short courses at different universities and conferences.

Scheduling a sports tournament involves some fundamental issues, such as the type of league, the tournament design, the timing of the scheduling, and the goals and constraints of each problem. Chapter 1 introduces some of these aspects, which will be examined in detail in the following chapters.

Chapter 2 describes the main combinatorial structures used to model, design, and optimize the scheduling of round-robin sports tournaments. Some fundamental problems are also introduced.

Chapter 3 focuses on local search, neighborhoods, and metaheuristics. A comprehensive study of edge-coloring neighborhoods used in local search and metaheuristics for scheduling round-robin sports tournaments is developed.

Chapter 4 illustrates how integer programming can be used to model and solve a round-robin tournament scheduling problem. The traveling tournament problem with predefined venues is used as the motivation for studying and comparing integer programming formulations of round-robin tournament scheduling problems.

Chapter 5 describes three applications of scheduling sports tournaments using approximate and exact methods described in the preceding chapters. We first discuss the minimization of weighted carry-over effects in round-robin tournaments. Following, we consider in detail the scheduling of the Brazilian annual football tournament. The third application reports a strategy for the dynamic scheduling of electronic sports tournaments, validated with real-life data.

Each chapter concludes with a detailed account of its related literature.

After revisiting the content of all chapters, we are happy to see that our goals have been attained. We are glad to realize that the study and development of combinatorial models and algorithms for scheduling sports tournaments have been a fantastic illustration of the beauty, usefulness, and applicability of Operations Research.

Due to our continued involvement with sports, most tournament scheduling problems are easy to explain and understand by researchers, developers, and practitioners, even if they may be tough to solve in computational terms. In addition, sports applications motivate students and can be used successfully in Operations Research courses and classrooms.

Writing this book was undoubtedly a challenging task. Above all, it has been a rewarding experience. We are delighted with the human interactions made possible by this project. We are very proud of its outcome and with the book coming to the shelves and hands of the readers. We will be happier if the readers and students appreciate reading and using it.

Furthermore, we will be even happier if the readers and students think about the problems and issues discussed in this book the next time they attend or watch a game of a fascinating, attractive, fair, and competitive tournament.

Have a good game, and enjoy the reading!

Rio de Janeiro, Brazil — Celso C. Ribeiro
Molde, Norway — Sebastián Urrutia
Lausanne, Switzerland — Dominique de Werra
December 2022

Acknowledgments

Over the years, we have collaborated with many coauthors on research related to topics covered in this book. The first two authors acknowledge their collaborators below, in alphabetical order, and apologize in case someone is missing in this list: Aleteia Araujo, Fabrício Costa, Zhi-Long Dong, Alexandre Duarte, Guillermo Durán, Allison Guedes, Edward Haeusler, Tiago Januario, Kui Jing, Graham Kendall, Sigrid Knust, Yujie Ma, Rafael Melo, Thiago Noronha, Kimmo Nurmi, Sebastian Souyris, Andrés Weintraub, Fengmin Xu, and Ailec Zamora. The third author thanks D. Costa, T. Ekim, A. Geinoz, and C. Raess, who have actively worked in sports scheduling. Their interest in this domain has led them to original contributions to the area.

The first author thanks Nelson Maculan for his friendship and support, particularly in recent years while this book was in preparation. The second author thanks Anolan Milanés and Paula Trillos for their daily support.

We acknowledge Mauricio Resende for the permission for reproduction of Figs. 3.6, 3.7, 3.8, and 3.9 that initially appeared in the book *Optimization by GRASP* (Springer, 2016), coauthored with the first author, David Van Bulck for providing information for the first draft of Table 5.9, and Carlos R. H. Márquez for his collaboration in the preparation of some figures.

We thank Christian Rauscher, Editorial Director at Springer, for his support and encouragement during the preparation of this book.

We acknowledge Thiago Jaconianni for creating the amazing illustration on page xxv that symbolizes this book. Just as sports scheduling is the focus of this work, it perfectly represents that all sports considered in our case studies and applications revolve around a ball.

We also acknowledge Rafael Melo's careful revision and insightful suggestions that improved our original text.

We are very grateful and indebted to Michael Trick for kindly accepting our invitation to write the foreword for this book, introducing our work and the field of sports scheduling to the readers.

Finally, we express our gratitude to many Operations Research and Computer Science students who discovered that sports scheduling is an inexhaustible source of exciting combinatorial optimization problems. Their questions and observations have been a strong motivation to work in this domain.

Rio de Janeiro, Brazil
Molde, Norway
Lausanne, Switzerland
December 2022

Celso C. Ribeiro
Sebastián Urrutia
Dominique de Werra

Contents

List of Abbreviations and Acronyms

BRKGA	Biased random-key genetic algorithm (metaheuristic)
CBF	Confederação Brasileira de Futebol (league)
CP	Constraint programming (method)
DRR	Double round robin (tournament design)
FBTS	First-break, then-schedule (decomposition method)
FSTB	First-schedule, then-break (decomposition method)
GA	Genetic algorithm (metaheuristic)
GPTS	Generalized partial team swap (neighborhood)
GR	Game rotation (ejection chain)
GRASP	Greedy randomized adaptive search procedures (metaheuristic)
HAA	Home-away assignment (array)
HAP	Home-away pattern (array)
HAS	Home-away swap (neighborhood)
ILS	Iterated local search (metaheuristic)
ITC	International Timetabling Competition
LS	Local search (method)
MA	Memetic algorithm (metaheuristic)
MIP	Mixed integer programming (method)
MINLP	Mixed integer nonlinear programming (method)
MLB	Major League Baseball (league)
mDRR	Mirrored double round robin (tournament design)
mTTP	Mirrored traveling tournament problem (problem)
NFL	National Football League (league)
NHL	National Hockey League (league)
NP	Non-deterministic polynomial (complexity)
PRS	Partial round swap (neighborhood)
PTS	Partial team swap (neighborhood)
RKGA	Random-key genetic algorithm (metaheuristic)
RS	Round swap (neighborhood)
SA	Simulated annealing (metaheuristic)
SAT	Satisfiability (problem)

SRR	Single round robin (tournament design)
TS	Tabu search (metaheuristic)
TS	Team swap (neighborhood)
TTP	Traveling tournament problem (problem)
TTPPV	Traveling tournament problem with predefined venues (problem)
VND	Variable neighborhood descent (local search method)
VNS	Variable neighborhood search (metaheuristic)

List of Algorithms

List of Figures

List of Tables

MAY THE BALL BE WITH YOU

Chapter 1
Leagues, Tournaments, and Schedules

The scheduling of a sports tournament involves some fundamental issues, such as the type of league, the tournament characteristics, the timing of the scheduling process, and the goals and constraints of each specific problem.

Sports teams and leagues want to optimize their investments by playing a good schedule that meets various criteria. Good schedules play a vital role in a sports tournament. They are important to maximize fairness and revenues, ensure the games' attractiveness, keep the interest of both the media and the fans, and make up for better logistics while reducing the organizers' costs. They can have significant financial implications and interfere in the performance of every team and player participating in a tournament.

The main problem in sports scheduling consists in determining the date (or the round), time, and venue (facility, stadium, court, or arena) in which each game of a tournament will be played. Applications exist in real-life scheduling tournaments of sports such as football, baseball, basketball, cricket, and hockey. Different exact and approximate approaches, including integer programming, constraint programming, metaheuristics, and hybrid methods, have been applied to these problems.

In this chapter, we introduce and discuss some of these aspects, which will be examined in more detail in the following chapters. Section 1.1 discusses different aspects of sports tournaments that impact their scheduling. Section 1.2 considers the characteristics of sports scheduling problems in professional and amateur leagues. Section 1.3 presents the main tournament scheduling problems in practice and literature. Section 1.4 describes the organization of the book and each chapter's content.

C. C. Ribeiro et al., *Combinatorial Models for Scheduling Sports Tournaments*,
EURO Advanced Tutorials on Operational Research,
https://doi.org/10.1007/978-3-031-37283-4_1

1.1 Tournament Characteristics

This section discusses different characteristics of sports tournaments that impact the scheduling. In particular, we consider the tournaments' compactness, timing, and design (or format).

1.1.1 Compactness

Tournament scheduling problems in sports leagues may be divided into two main classes: time-constrained and time-relaxed problems.

The horizon of a *time-constrained tournament* consists of the number of *rounds* (periods, days, or time slots) required to schedule all games. Each team plays, at most, one game in each round. The rounds are synchronized, one after the other. The games of round $k + 1$ are played after the games of the previous round k have been played. Time-constrained tournaments that, in addition, impose a minimum number of rounds so that every team plays once in each round are said to be compact. Most professional football leagues in Europe and South America use compact, time-constrained tournaments.

There are no synchronous rounds in a *time-relaxed tournament*. Thus, some teams can stay long without playing a game. In contrast, other teams may play multiple games in a row (this is the typical situation of leagues where some teams perform long trips when they play several consecutive away games without returning home). Time-relaxed tournaments offer more flexibility to handle the limited availability of teams and venues. The horizon of a time-relaxed tournament is much longer than necessary, resulting in more flexibility and fewer infeasibility issues. The professional tournaments promoted by the National Basketball Association (NBA) and the National Hockey League (NHL) in North America are two examples.

1.1.2 Scheduling Timing

Another aspect involved is the timing at which the scheduling of the tournament is performed.

In the case of *static scheduling*, the schedule is determined once and for all before the tournament. The rounds, dates, and venues where the games will be played are known beforehand, with simplified logistics issues since the league, the teams, the press, the fans, and all other participants can plan well in advance. This case is typical of the annual tournament of national professional football leagues, where each team plays against every other once or twice, and every team plays exactly once in every round. The dates and stadiums of all games are known in advance.

Regarding *dynamic scheduling*, the schedule is determined round by round. The games in round $k+1$ (and their schedule) are determined after round k is completed, and all competitors' rankings are updated. A typical example is tennis tournaments, where players are seeded initially according to their ranking in the season. The players in the next round's games and the courts where they will play are defined after the winners of the games in the current round become known. Most chess tournaments also follow a dynamic scheme, with the competitors pairing based on their number of points (a player with two draws can be matched with a player with one victory and one loss). Some national football cups also follow this format. We remark that some tournaments follow a hybrid format scheduling, combining static and dynamic scheduling phases to increase their attractiveness. In the past, the Brazilian annual football tournament used this strategy for many years.

1.1.3 Design

Different designs exist for a sports contest or tournament. A tournament's design (or format) may influence its outcome uncertainty, the number of unimportant matches within the tournament, and even the fairness of the tournament. Low uncertainty of outcome may mean that fan, broadcaster, and sponsor interest will not be maximized. This section discusses three basic tournament designs (round robin, knockout, and Swiss system), with all other designs considered their variations and hybrids.

1.1.3.1 Round-Robin Tournaments

A *round-robin tournament* (or *all-play-all tournament*) between an even number of $2n$ competitors is one in which each competitor plays precisely the same number of games with every other competitor in the tournament. This term is not used when one contestant plays others an unequal number of times, as in almost all of the major United States professional sports leagues.

Most professional sports leagues play either a single round-robin tournament (SRR), where each pair of teams meets precisely once, or a double round-robin tournament (DRR), where each pair of teams meets twice. Generally, each competitor plays against every other competitor exactly m times. The number of rounds in a compact tournament (i.e., a tournament where each competitor plays precisely once in each round) equals $m(2n-1)$. Double round-robin tournaments can be organized in different manners, such as unconstrained tournaments in which the games can be played in any arbitrary order; tournaments with constraints avoiding that the two games between the same pair of contestants be played in two consecutive rounds; tournaments organized in two separate phases, each of which is an SRR tournament (possibly with different timetables); and mirrored tournaments (both phases are SRRs with the same timetables, but the venues of the two games between each pair of competitors are interchanged from one phase to the other).

Double round-robin tournaments are standard in leagues with many competitive games. Frequently, preliminary group phases within a wider tournament are conducted as a single round-robin tournament, such as the FIFA World Cup and the UEFA European Football Championship. Most tournaments of football leagues are organized with the double round-robin design, in which every team plays all others in its league once at home and once away.

In principle, a round-robin tournament would be the fairest way to determine the champion or winner from among a known and fixed number of competitors. Each of them has equal chances against all other competitors. No initial seeding precludes a game between any given pair of competitors. Every competitor plays against every other, and competitors are not eliminated after a certain number of losses. Competitors accumulate points along the tournament depending on their results (wins, draws, or losses) and, in some cases, their game scores.

However, round-robin tournaments have many more games compared to other tournament designs and, in consequence, can be very long, limiting the number of competitors in such tournaments. Professional leagues whose competitions are played as round-robin tournaments are usually limited to a few more than 20 competitors.

Furthermore, round-robin tournaments can suffer from games scheduled in the last rounds potentially not having any substantial importance. Since the tournament winner gradually emerges along multiple rounds of play, competitors who perform poorly are forced to play out their remaining games. Some later matches will pair one competitor with something left to play for against another who does not, which also opens the door to easier manipulations of results. Many games may be uninteresting due to the imbalance between the opponents, and no effective strategy exists to maximize game-by-game tournament attractiveness. It may also occur that some competitors play sequences of consecutive games against strong opponents while others play easy sequences, creating imbalance and showing that playing the same opponents is not necessarily completely equitable. Schedules where stronger and weaker opponents are evenly spreaded over the tournament duration for all teams are more balanced. This situation leads to the carry-over effects value problem, which will be introduced in Sect. 1.3.5 and studied in Sect. 5.1.

1.1.3.2 Knockout or Single-Elimination Tournaments

A *knockout tournament*, also called a *single-elimination tournament*, is an elimination tournament between $2n$ competitors (here, $2n$ is a power of two, i.e., $2n = 2^\ell$ for some integer $\ell \geq 1$) where the loser of each game is immediately eliminated from the contention for the title of champion. Each winner will play another winner in the next round until the final match-up, whose winner becomes the tournament champion. Variants include double- or triple-elimination or a combination with a best-of-k series, when a competitor is eliminated after losing $(k + 1)/2$ out of k games with the same opponent, where k is odd.

Some real-life competitions are held with a pure single-elimination tournament format. This design is typical, for example, of national football cups. It is also the typical case of Grand Slam tennis tournaments, in which the players are seeded in different brackets according to their current rankings. If the best-ranked opponent wins every game, the final match-up will see the two best-ranked players face to face. Other competitions may have many phases, with the last being a single-elimination final stage, often called playoffs. The FIFA World Cup is one of the most noticeable sports events organized with this structure. In the first phase, the national football teams are organized into groups of four teams that play all-to-all. The two best-ranked teams in each group proceed to the second phase, a knockout tournament.

The single-elimination format allows a much larger number of competitors to participate in the tournament than in the round-robin format. Every game counts, and there are no games where one competitor has more to play for than the other: every game is equally essential for both competitors, the winner gets all, and the loser is eliminated. We observe that this format is unsuitable for sports where games may end up with a draw. Most competitors are eliminated after relatively few games: one-half of the competitors still in contention is eliminated at the end of each round. Variants such as the double-elimination format are used tentatively to overcome this disadvantage.

Opponents may be allocated randomly, but this may result in the highest-rated competitors being scheduled to face each other early in the competition. Seeding is often used to prevent this. Brackets are set up so that the top two seeds can not possibly meet until the final round, none of the top four can meet prior to the semifinals, and so on. Other seeding strategies may be used. Seeding and bracketing are very relevant to the outcome of the tournament. For a given seeding and bracketing strategy, the games to be played in the first round are directly set. The results of the current round automatically determine the games to be played in the next. Basically, seeding and bracketing define the schedule.

The knockout format is also known as a *sudden death tournament*.

1.1.3.3 Swiss-System Tournaments

A *Swiss-system tournament* between $2n$ competitors is a non-eliminating format featuring a fixed number of rounds of competition, which is considerably smaller than the number of rounds of a single round-robin tournament. Competitors, either a team or an individual, do not play against all others. They are paired in each round using a set of rules designed to ensure that each competitor plays an opponent with a similar score (or ranking) but does not play the same opponent more than once. The winner is the competitor with the best aggregate score (or ranking) in all rounds. Since the number of competitors is even, they all play in every round.

Table 1.1 Swiss-system tournament played by eight teams: assuming team A wins the game between A and B in the third round, it will necessarily face a team with fewer wins in the next round

	Teams							
Number of wins	A	B	C	D	E	F	G	H
After first round	1	1	1	1	0	0	0	0
After second round	2	2	1	1	1	1	0	0
After third round	3	2	1 or 2	1 or 2	1 or 2	1 or 2	0 or 1	0 or 1

The Swiss system is used for competitions where there are too many competitors for an entire round-robin tournament to be feasible in practice due to timing constraints, and eliminating any competitor before the end of the tournament is undesirable. The Swiss-system design was first used in a chess tournament in Zurich in 1895, which is how it earned its name. Unlike designs like round robin, where all pairings are known beforehand, the scheduling is dynamic in a Swiss-system tournament. Game pairing for the next round is performed after the previous round has ended and depends on its results. The Swiss system seeks to provide a clear winner from many competitors in a relatively small number of rounds of competition without a single bad result terminating participation.

In all rounds except the first, competitors are paired based on their performance until this point of the competition. In the first round, they are paired either randomly or according to some pattern. If it is desired for top-ranked participants to meet in the last rounds, the pattern must place them in different games, similar to the seeding in brackets of pre-ranked players for knockout tournaments. In subsequent rounds, competitors are paired with opponents with a similar cumulative score (or rating) up to that point. The pairing rules may be quite complicated, as they have to ensure that no two competitors ever oppose each other twice and avoid giving a player some advantage due to chance.

As a short example, we consider a tournament with eight teams, as illustrated in Table 1.1: A, B, C, D, E, F, G, and H. Suppose the four games A vs. E, B vs. F, C vs. G, and D vs. H take place in the first round. The winners are A, B, C, and D. Since opponents with similar rankings are paired in each round, the second round could be formed, e.g., by games A vs. C, B vs. D, E vs. G, and F vs. H. Again, suppose the winners are A, B, E, and F. Then, at the end of the second round, A and B have two wins each, C to F have one win each, and G and H have no wins. The third round will necessarily see the match between teams A and B, and after they meet, there will be only one team leading the tournament with three wins. Similarly, only one team will have no wins (G or H). In the next round, the winner of the game between A and B will have three wins and will face a lower-score team.

The Swiss system and its many variants have been used for tournaments of many sports, including chess, go, bridge, badminton, curling, and electronic sports.

Swiss-system tournaments attempt to combine elements of the round-robin and elimination formats to provide an outcome using fewer rounds than a round robin while allowing draws and losses.

1.2 Professional Versus Amateur Leagues

Sports have become a big business. Teams make investments in new players. Broadcast rights amount to hundreds of millions of dollars in some competitions. In a globalized economy, many countries and cities battle for the right to organize major events such as the Olympic Games and the FIFA World Cup, which often bring thousands of jobs, urban regeneration, and economic opportunities to their hosts. Tournaments are followed by millions of people worldwide, eager for information on their teams' progress in each competition. Fans check newspapers, radio, television, and the Internet for information.

Professional sports leagues involve millions of fans and significant investments in players, broadcast rights, merchandising, and advertising. Multiple agents, such as the organizers, media, players, fans, security forces, and airlines, play an essential role in the leagues and tournaments. Professional sports leagues are, therefore, part of a major economic activity and face challenging optimization problems, such as game scheduling, revenue maximization, and logistic optimization. Electronic sports (or, simply, e-sports) bring new challenges due to their tournaments' different nature and design.

On the other hand, amateur leagues usually do not have access to the same resources. However, the number of tournaments and participants can be huge, requiring coordination and logistical efforts. The scheduling problems in such leagues may enormously differ from those of professional leagues. Amateur leagues of several sports, such as baseball, basketball, and football, have hundreds of games every weekend in different divisions of age and gender, while professional leagues usually have around 20 teams and fewer games.

In a single amateur league of a county in the United States, for example, there might be up to 500 football games in a weekend involving around 10,000 players and hundreds of referees. These games may belong to different league divisions based on the players' age, gender, or level. In an amateur league, one has to avoid long travel times on weekdays while assigning to the weekends the games that imply more extended travels for the teams, organizers, and referees.

Referee assignment is subject to many criteria. Due to the shortage of certified referees, each referee may officiate at several games on the same day. In case the referees can officiate in different facilities, some objectives may be minimizing the total number of inter-facility travels, the total traveling time, or the idle times between consecutive games assigned to the same referee. Tournament organizers of amateur leagues may also want referee assignments that match preferences regarding the facilities, divisions, and time slots where the referees officiate.

In addition to the problem of creating a reasonable timetable over the season and assigning the games to the venues, another problem in youth amateur leagues is practice assignment, in which teams must be assigned to off-school recurring practice time slots that are available through the season. Coaches express their preferences regarding time slots, to which preference weights and coaching constraints are applied. The optimization process attempts to maximize the overall weighted satisfaction while giving a minimum guaranteed satisfaction to all teams.

Team assignment is still another problem in amateur leagues. The candidate players participating in a league specify their preferences and requirements for team assignments. They can specify the following: the friends they want to play in the same team, the coach they prefer, or the team they want to join. Players are assigned to the teams, maximizing their satisfaction so that every team has a guaranteed minimum number of players and no more than a specific maximum limit.

Another challenging problem in amateur and youth sports is multi-league scheduling, where timetables for multiple leagues must be determined simultaneously. The leagues play round-robin tournaments with different numbers of teams, thus requiring different numbers of rounds. As the number of simultaneous home games that the clubs can organize for their teams is limited by the capacity of their venue, the objective is to minimize capacity violations.

1.3 Typical Problems

The Swiss system uses a dynamic scheduling strategy: the next round's games only become known after all those of the current round finish. A similar situation also occurs with knockout tournaments. The games to be played in the next round also depend on knowing the winners in the current round. These designs may impose additional logistics difficulties in the case of extramural tournaments where each team has a venue in its home city. Decisions associated with traveling and displacement of teams (as well as those related to security, media coverage, and displacement of the fans) will have to be made quickly on the fly.

Therefore, the Swiss-system and knockout designs are more appropriate for large tournaments held in a single place (such as a large tennis complex with many courts) or over the internet (as in the case of electronic sports). On the other hand, the round-robin format is more suited to professional competitions where some teams have to travel after a game to play the next, and all the logistics have to be prepared well in advance.

This book focuses mainly on the static scheduling of compact (time-constrained) round-robin tournaments, although other problems, such as the scheduling of electronic sports, are also discussed. Some of the main tournament scheduling problems are introduced in the following sections.

1.3.1 Break Minimization

In the case of extramural tournaments, each team has its stadium located in its home city. Every team plays at home against a visitor or away in another city as a visitor in each round. Playing a home game is more comfortable: players have more time and better conditions for resting, practicing, and recovering from the previous game. Furthermore, the players are closer to the fans and get much more support at game time. Playing an away game involves traveling and, sometimes, restless journeys.

Every team participating in the tournament should have a similar number of home and away games.

A sequence of home games is always advantageous for any team, while a road trip of away games is tiresome and often prejudicial. Fair schedules seek balanced sequences of home games and away games for every team.

An ideal strategy to maximize fairness would be to perfectly alternate home and away games for every team. Since this is impossible, we may search for a schedule that minimizes the breaks in the alternation of home and away games for all teams: this is the *break minimization problem.*

1.3.2 Distance Minimization

In some tournaments, the home cities of the participating teams may be dispersed unevenly over a large territory, and the distances between them may vary a wide range. Traveling from A to B may take much longer or cost substantially more than from A to C.

We assume every team begins the tournament at home and must return after its last away game. In this case, we are interested in scheduling the games in such a way that the total distance traveled by the teams (or the total cost of the displacement of the teams) from one round to the next is minimized, considering all rounds: this is the *distance minimization problem.*

Minimizing the total distance traveled (or cost) by all teams may lead to a severe imbalance in the distances traveled by each team. Although of minimum distance, the schedule may be unfair to some teams that concentrate most of the traveling and highly beneficial to others. Additional constraints have to be imposed to guarantee fairness or equilibrium.

1.3.3 Traveling Tournament Problem

Some problems are not concerned with an actual world application but have been introduced as fundamental problems that may be expanded and modified according to the needs of sports leagues. Their structure and simplicity stimulate the elaboration of solution techniques that may be used subsequently for real-world problems.

Perhaps the most valuable and emblematic of these is the *traveling tournament problem* (TTP), which aims to minimize the total distance traveled under some specific, realistic constraints. It provides simplified objectives and constraints of a real-world problem. It is a challenging optimization problem in sports scheduling that exhibits the most critical aspects in creating timetables where travel distances or costs are to be considered.

Given two integer numbers L and U, and a set of teams whose distances between their home cities are known, the problem calls for the schedule of a double round-robin tournament minimizing the total distance traveled by the teams and respecting a set of constraints. Every team begins the tournament at home and must return after its last away game. Whenever a team plays two consecutive away games, it goes directly from the site of the first opponent to that of the second without going back home. No repeaters are allowed, i.e., no two teams can face each other in two consecutive rounds. Furthermore, every sequence of consecutive home games (or consecutive away games) played by any team is formed by at least L and, at most, U games.

1.3.4 Cost Minimization

In a general model, one may assume that varying costs (or benefits) that depend on the round are associated with every game. In particular, for each ordered pair of teams i and j, suppose there is a cost c_{ijk} of scheduling the game between them as a home game for team i in round k. The objective is to find a feasible round-robin tournament schedule satisfying several constraints and with *minimum total cost* (or *maximum total benefit*).

This problem is similar to the maximization of the total satisfaction of the participating teams, which may express their wishes or preferences for playing some specific games at home or away or in some specific rounds.

1.3.5 Carry-Over Effects

A significant issue in a competitor's strategy, particularly in long competitions, consists of balancing their efforts over the competition. If a competitor plays against a weak opponent, it will likely be in better shape to play in the next round than if it had played against a hard one. Competitors who play against strong opponents will likely be more tired at their next game. Therefore, it is likely that a competitor would make much less effort playing against an opponent that played before against a very hard competitor than it would make against an opponent that faced an easy competitor.

The above situation is particularly true in the case of sports that require a significant amount of physical effort (such as wrestling or martial arts). In this sort of sport, it is not uncommon that a competitor plays several matches in a row, making a sequence of very tired (resp. well-reposed) opponents attractive (resp. unattractive). Some schedules may contain several sequences of easier (or harder) games in a row assigned to one or more competitors. This situation does not characterize a fair schedule and is highly undesirable in any tournament. To illustrate this effect, consider a karate or judo competition, for which there is no weight division in open-weight categories: a physically weak athlete may fight a strong one. A competitor who has just fought a powerful opponent might be tired (and even wounded) in the next fight. A performance deterioration could give the next opponent a decisive advantage.

We consider that competitors play a single round-robin tournament. We say that competitor C receives a *carry-over effect* due to B if competitor A plays C just after its match against B. Thus, one may count the value of each competitor's carry-over effect to every other in the tournament schedule.

A fair tournament regarding the value of the carry-over effects is one in which the latter are evenly distributed over all pairs of teams, and the total value of the carry-over effects is minimized: this is the *carry-over effects minimization problem.*

Carry-overs determined by one specific team (and even a player) strongly affected the outcome of the main division tournaments of some European leagues. Another interesting real-life situation was observed in US college football, whereby a team was repeatedly scheduled against teams with byes the week before. The sequence of games could have been more attractive for this specific team because it was scheduled to meet frequently a well-rested team that had not played in the previous round. In fact, in tournaments with an odd number of competitors, it is customary to create a dummy (or artificial) team and to produce the schedule for an even number of teams. Whenever a team faces the dummy team, it has a bye in this round. In such situations, no team must have many games against teams that had a bye in the previous round to avoid unfair schedules.

1.3.6 Referee Assignment and Umpire Scheduling

The assignment of referees (or crews of referees) to the already scheduled games of a tournament is the most studied scheduling problem in sports after tournament scheduling. Applications to professional sports leagues such as baseball, cricket, football, and tennis appear in the literature. Fairness issues, such as avoiding frequent assignments of the same referee to the games of the same team, are among the most critical constraints and motivations for these applications.

The referee assignment problem is also common in many amateur leagues of different sports, such as football, baseball, and basketball. In this problem, a limited number of referees with different qualifications and availabilities should be assigned to a set of already scheduled games. The number of referees in a crew may vary, depending on the sport, league, and tournament: football games usually require three referees on the field, while basketball games require two. Typical applications involve youth leagues where hundreds of referees should be assigned to hundreds of games in a short period, such as a weekend. Some of the requirements to be satisfied by the referee assignments are: (1) referees can not officiate at more than one game in overlapping time slots or in time slots where they declared to be unavailable; (2) a fair dispersion of referees to games throughout the season, (3) the number of times a referee is assigned a team when it is playing at home, (4) higher-skilled referees should be assigned to more games than less qualified referees, (5) referees must meet the minimum skill level established for each game and division, (6) the same referee should not officiate at both games between the same pair of teams in the case of double round-robin tournaments, (7) each referee should be allocated to a number of games as close as possible to an agreed target number (which might be different for each referee), and (8) game assignments in the same facility and day should be privileged to reduce travel times.

The assignment of judges in some non-sport competitions, such as the John Molson MBA International Case Competition, has some features in common with the assignment of referees in sports leagues. Senior professionals with relevant business and management experience from various backgrounds are welcomed as judges.

1.3.7 Multileague Scheduling

Schedules for professional and amateur tournaments, in some contexts, should consider the inter-relation among several tournaments in which the teams of a league participate. In some cases, the schedule maker must be aware of the schedules of other leagues, and some coordination may benefit both.

In professional sports, teams often play more than one tournament simultaneously. In a typical situation, a team plays both national and international tournaments at the same time. Ideally, the schedule of both competitions should be done simultaneously and taking into account each other. However, a hierarchical approach is usually applied in which the procedure for constructing the national tournament receives the already available schedule of the international tournament as input.

A similar situation is encountered when scheduling different divisions of a league. The games of one division share resources with other divisions, and ideally, all their schedules should be prepared together. In practice, the schedule of the lower division is done after the one of the upper league is finalized.

In addition, the schedules for national and international competitions are subject to the calendars of the competitions of the national teams since the players are shared resources among teams and national teams.

In amateur sports competitions, coordinating the schedules of several divisions and sports is even more challenging. Not only there is no clear hierarchy among divisions (e.g., by age, gender, and sport discipline), but there are also many diverse shared resources such as fields, time slots, referees, hardware, and even players. Furthermore, the number of teams and players involved in such amateur competitions is much larger than in the case of professional sports. The number of divisions sharing resources may reach the hundreds, while the number of players involved may be in the thousands; see Sect. 1.3.6 for shared resources for the referee assignment problem.

The coordination of the schedules of several leagues becomes essential in this context. For example, consider the problem of determining the starting date of each competition to avoid having teams from too many leagues using the same facility at the same time.

1.3.8 International Timetabling Competition

Sports scheduling was the theme of the Fifth International Timetabling Competition (ITC 2021). Its goal was to promote and provide insights into the development of more generally applicable sports timetabling solvers, considering that most works in the literature dealt with particular sports scheduling problems applied to a specific context (in terms of the league, sport, and region).

The problem asks for a schedule for a compact double round-robin tournament with 16, 18, or 20 teams. In some instances, the schedule must be phased, meaning that it should be composed of two consecutive single round-robin tournaments with an opposite home-away assignment.

The constraints are separated into soft and hard and divided into five categories. Capacity constraints force a team to play home or away in a given round, limit the number of games a team plays against a particular group of teams in a given period, set a limit on the number of consecutive home or away games, or enforce complementary patterns of home and away games. Game constraints enforce or forbid specific assignments of games to rounds. Fairness constraints impose that the maximum difference between the number of home games played by any pair of teams is, at most, two. Break constraints regulate the maximum number and timing of breaks. Separation constraints impose a lower bound on the number of rounds separating two games between the same pair of teams.

To encompass as many real problems as possible, the soft and hard constraints vary from one instance to another. The objective function minimizes the penalties associated with the violated soft constraints, while the hard constraints must necessarily be met.

1.4 Book Organization

This book has four additional chapters.

Chapter 2 describes the main combinatorial structures used to model, design, and optimize the scheduling of round-robin sports tournaments.

Chapter 3 focuses on local search, neighborhoods, and metaheuristics. A comprehensive study of solution neighborhoods used in metaheuristics for scheduling round-robin sports tournaments is developed.

Chapter 4 illustrates how integer programming can be used to formulate and solve a typical round-robin tournament scheduling problem. The traveling tournament problem with predefined venues is used as the motivation for the study of integer programming formulations of round-robin scheduling problems.

Chapter 5 describes three case studies of tournament scheduling applications in detail, using some of the approximate and exact methods described in the preceding chapters.

Each chapter concludes with a detailed account of its related literature.

1.5 Related Literature

Section 1.1 presents and discusses the main characteristics of sports tournaments. Scarf et al. (2009) observed that a tournament's design might influence its outcome uncertainty and fairness. Owen and Weatherston (2004) suggested that low uncertainty of outcome may mean that the interest in the competition will not be maximized.

The book by Rokosz (2000) provides a basic introduction to several types of tournaments, such as single elimination, double elimination, round robin, and others.

Harary and Moser (1966) discussed the theory of round-robin tournaments. Suksompong (2016) studied the problem of scheduling asynchronous round-robin tournaments. Van Bulck and Goossens (2022b) reviewed the literature on time-relaxed tournaments, which offer more flexibility to handle the limited availability of teams and venues, and proposed an iterative two-phase approach to construct non-compact round-robin timetables. Hwang (1982), Schwenk (2000), and Marchand (2002) discussed and compared seeding strategies for knockout tournaments. According to Scimia (2019), the Swiss system was first employed at a chess tournament in Zurich in 1895, which is how it earned its name, and is now used in many sports, including electronic sports, chess, go, and bridge. Ranking strategies for Swiss-system tournaments are described, e.g., in Csatò (2013, 2017) and Führlich et al. (2021). The Swiss system is considered by some authors the most accurate among all tournament formats, especially in its ability to rank all participants: it does not eliminate a competitor after a single loss, while it takes the history of the comparisons into account (Sziklai et al., 2022).

Dinitz et al. (2007) provided an introduction to some of the fundamental issues in sports tournament scheduling, including round-robin tournaments, 1-factorizations, and balanced tournament designs. The book of Anderson (1997) covers many of the basics that are of interest to sports scheduling, including Latin squares, Room squares, scheduling leagues, balanced tournament designs, and whist tournaments (Julian et al., 2003; Finizio and Mosconi, 2010).

Easton et al. (2004), Rasmussen and Trick (2008), Kendall et al. (2010), and Ribeiro (2012) presented surveys of sports scheduling problems. Knust (2021) hosts a classification of the literature on sports scheduling. Van Bulck et al.

(2020) described RobinX, a three-field classification and unified data format for round-robin sports timetabling, whose associate repository with all collected data (test instances, bounds, optimal solutions) and software tools (such as a solution validator) is available on the website devoted to this project (Van Bulck et al., 2021b).

Section 1.3 gives an introductory description of the principal, typical sports tournament scheduling problems. The first objective in scheduling sports tournaments is the minimization of breaks. Many league organizers restrict themselves to schedules with minimum breaks; see de Werra (1980, 1981, 1988) for pioneering work on the subject. Suzuka et al. (2007) presented a unified view of the home-away assignment problem and the break minimization problem. Regin (2001) presented constraint programming approaches for break minimization in single round-robin tournament problems. Solutions for the (polynomially solvable) problem of finding a schedule with $n - 2$ breaks for instances with $n = 20$ teams were produced in 0.61 seconds and in less than a minute for 60 team instances. Afterward, some experiments were done for the much more difficult situation when a timetable (i.e., an opponent schedule) is given, and breaks are to be minimized. Instances with up to 12 teams could be solved by introducing additional global constraints in the constraint programming formulation. Urrutia and Ribeiro (2006) showed that sometimes a large number of breaks is advantageous since, in this case, travel distances are minimized in the traveling tournament problem.

The timetable-constrained distance minimization problem was applied by Rasmussen and Trick (2006) for tournaments where the total travel distance must be minimized. A hybrid approach combining integer programming and constraint programming was proposed for its solution.

The traveling tournament problem (TTP) was introduced by Easton et al. (2001). The mirrored traveling tournament problem was tackled in Ribeiro and Urrutia (2007). The traveling tournament problem with predefined venues is a single round-robin variant of the traveling tournament problem described in Melo et al. (2009). Different exact and approximate solution methods have been applied to the TTP and its variants. A combined integer programming and constraint programming approach for the TTP was proposed by Easton et al. (2003), where the first provably exact solution for a TTP instance with eight teams was presented, although the no-repeaters constraint has been discarded.

Briskorn and Drexl (2009a,b) developed algorithms to schedule a round-robin tournament with the minimum number of breaks and the minimum total cost. Briskorn and Horbach (2012) employed a Lagrangean relaxation approach to obtain tight lower bounds for minimum-cost single round-robin tournaments. They also developed a cost-oriented repair mechanism yielding a feasible tournament schedule for each solution of the relaxed problem. Van Bulck and Goossens (2022a) explored how to generalize decomposition techniques commonly used for compact schedules to generate time-relaxed timetables less prone to some drawbacks. Among such drawbacks, we mention the variability of the rest periods between competitors' consecutive games or significant differences between the number of games played at any time of the season between two competitors. Yi and Goossens (2023) focused on dealing with uncertain events in time-relaxed round-robin sports timetabling and how to mitigate their effect on the quality of the timetable. Proactive strategies determine the position of time slots used as buffers, while reactive strategies reschedule matches to these buffers considering three quality measures. The objective of Knust and Lücking (2009) was to find a feasible schedule for a round-robin tournament with the minimum number of breaks and the minimum total cost, where additional place constraints are taken into account.

The system developed by Bach et al. (2018) to schedule the elite football series in Norway is based on the maximization of the teams' wishes.

Anderson (1999) developed a construction scheme for schedules with small carry-over effects values. In Miyashiro and Matsui (2006), a simple heuristic was proposed to quickly generate schedules with small carry-over effect values. The heuristic makes use of the circle method for constructing canonical 1-factorizations associated with schedules with high carry-over effect values. Permutations of the rounds of the schedules generated by the circle method give better solutions with small carry-over effect values. A heuristic based on the iterated local search metaheuristic was proposed by Cao et al. (2022).

Earlier attempts to apply optimization techniques to umpire scheduling in professional baseball leagues appeared in Evans et al. (1984), Evans (1988), and Ordonez and Knowles (1998), as well as in professional leagues of tennis (Farmer et al., 2007). Yavuz et al. (2008) presented an integer programming model for the fair assignment of referees to football games, avoiding frequent assignments of the same referee to games of the same team. A constructive heuristic and a local search procedure were developed for its solution.

The Major League Baseball (MLB) comprises 30 teams that play 2430 games annually during a six-month season. Umpires travel from city to city because one of MLB's objectives is to ensure that an umpire does not handle the games of any team too frequently in one season. Although minimizing the travel of umpires is essential, the requirement that a crew does not handle the games of any team too often forces each crew to travel after each series of games. The crews must travel long distances during a season because they must visit each MLB home city, and there are restrictions on a crew handling consecutive series for any team. The approach developed by Trick et al. (2012) to schedule MLB umpires is based on network optimization and simulated annealing. The method was used to schedule MLB umpires in 2006 and 2008–2010; see also Xue et al. (2015), Toffolo et al. (2016), Oliveira et al. (2016), and Chandrasekharan et al. (2019) for the traveling umpire problem.

Duarte et al. (2007b) introduced the referee assignment problem in youth football amateur leagues. A three-phase heuristic approach was proposed based on a constructive procedure, a repair heuristic to make solutions feasible, and an ILS improvement heuristic. Duarte et al. (2007a) extended and improved the three-phase approach. A new construction algorithm based on a greedy criterion was proposed to build low-cost initial solutions. Also proposed was an innovative hybridization strategy in which a mixed integer programming exact algorithm replaced the original neighborhood-based local search within the ILS heuristic.

Duarte and Ribeiro (2008) considered a multi-objective version of the previous problem and developed a heuristic to find good approximations of the Pareto set for real-size instances, based on extensions of the three-phase strategy initially proposed for the single objective problem version.

The multileague scheduling problem was studied by Davari et al. (2020) and Li et al. (2022). The first article assumed that all leagues have the same size, and a polynomial-time algorithm was developed. The second article dealt with leagues of different sizes.

The Fifth International Timetabling Competition (ITC 2021) was described in Van Bulck et al. (2021a), which includes the problem descriptions, the competition rules, and the final results. The test instances were made available on the website developed by Van Bulck et al. (2021), using a standardized XML data format (Van Bulck et al., 2020, 2021b), together with a software tool for solution validation. The competitors could use any commercial solver, and no time limit was imposed for the solution methods.

Lamghari and Ferland (2005, 2007, 2010, 2011) developed and compared heuristics and metaheuristics to assign the judges for the John Molson International Case Competition. The competition is a round-robin tournament with seven business cases open to top business schools worldwide (John Molson, 2020).

Due to our daily and continuous involvement with sports, most tournament scheduling problems are easy to explain and understand by researchers, developers, and practitioners, even if they may be tough to solve in computational terms. Sports applications motivate students and can be successfully used in Operations Research courses and classrooms, as illustrated in the experiences reported by Trick (2004) and Goossens and Beliën (2022).

References

Anderson I (1997) Combinatorial designs and tournaments. Oxford lecture series in mathematics and its applications. Clarendon Press, Oxford

Anderson I (1999) Balancing carry-over effects in tournaments. In Holroyd F, Quinn K, Rowley C, Webb B (eds) Combinatorial Designs and Their Applications. Chapman & Hall, London, pp 1–16

Bach L, Kloster O, Ribeiro CC (2018) Scheduling professional football leagues. In: Conference handbook of the 29th european conference on operational research, p 398

Briskorn D, Drexl A (2009a) A branch-and-price algorithm for scheduling sport leagues. J Oper Res Soc 60:84–93

Briskorn D, Drexl A (2009b) A branching scheme for finding cost-minimal round robin tournaments. Eur J Oper Res 197:68–76

Briskorn D, Horbach A (2012) A Lagrangian approach for minimum cost single round robin tournaments. Comput Oper Res 39:718–727

Cao Y, Wu W, Yagiura M (2022) A metaheuristic algorithm for the carry-over effect value minimization problem in round-robin tournaments. J Adv Mech Des Syst Manufact 16(4):JAMDSM0042

Chandrasekharan RC, Toffolo TAM, Wauters T (2019) Analysis of a constructive matheuristic for the traveling umpire problem. J Quant Anal Sports 15:41–57

Csatò L (2013) Ranking by pairwise comparisons for Swiss-system tournaments. Central Eur J Oper Res 21:783–803
Csatò L (2017) On the ranking of a Swiss system chess team tournament. Ann Oper Res 254:417–436
Davari M, Goossens DR, Beliën J, Lambers R, Spieksma FCR (2020) The multi-league sports scheduling problem, or how to schedule thousands of matches. Oper Res Lett 48:180–187
Dinitz JH, Fronček D, Lamken ER, Wallis WD (2007) Scheduling a tournament. In: Colbourn CJ, Dinitz JH (eds) Handbook of combinatorial designs, 2nd edn. CRC Press, Boca Raton, pp 591–606
Duarte AR, Ribeiro CC (2008) Referee assignment in sports leagues: approximate and exact multi-objective aproaches. In 19th international conference on multiple criteria decision making, pp 58–60
Duarte AR, Ribeiro CC, Urrutia S (2007a) A hybrid ILS heuristic to the referee assignment problem with an embedded MIP strategy. In: Bartz-Beielstein T, Aguilera MJB, Blum C, Naujoks B, Roli A, Rudolph G, Sampels M (eds) Hybrid metaheuristics. Lecture notes in computer science, vol 4771. Springer, Berlin, pp 82–95
Duarte AR, Ribeiro CC, Urrutia S, Haeusler EH (2007b) Referee assignment in sports leagues. In: Burke EK, Rudová H (eds) Practice and theory of automated timetabling VI. Lecture notes in computer science, vol 3867. Springer, Berlin, pp 158–173
Easton K, Nemhauser G, Trick MA (2001) The traveling tournament problem: description and benchmarks. In: Walsh T (ed) Principles and practice of constraint programming. Lecture notes in computer science, vol 2239. Springer, Berlin, pp 580–585
Easton K, Nemhauser GL, Trick MA (2003) Solving the travelling tournament problem: a combined integer programming and constraint programming approach. In: Burke EK, de Causmaecker P (eds) Practice and theory of automatedtimetabling IV. Lecture notes in computer science, vol 2740. Springer, Berlin, pp 100–109
Easton K, Nemhauser GL, Trick MA (2004) Sports scheduling. In: Leung JT (ed) Handbook of scheduling. CRC Press, Boca Raton, pp 52.1–52.19
Evans JR (1988) A microcomputer-based decision support system for scheduling umpires in the American Baseball League. Interfaces 18:42–51
Evans JR, Hebert JE, Deckro RF (1984) Play ball - the scheduling of sports officials. Perspect Comput 4:18–29
Farmer A, Smith JS, Miller LT (2007) Scheduling umpire crews for professional tennis tournaments. Interfaces 37:187–196
Führlich P, Cseh A, Lenzner P (2021) Improving ranking quality and fairness in Swiss-system chess tournaments. arXiv:2112.10522v1. Retrieved January 30, 2023
Finizio NJ, Mosconi SJN (2010) Balanced whist tournaments. J Combin Math Combin Comput 73:143–158
Goossens DR, Beliën J (2022) Teaching integer programming by scheduling the Belgian soccer league. INFORMS Trans Edu 23:137–217. https://doi.org/10.1287/ited.2022.0269
Harary F, Moser L (1966) The theory of round robin tournaments. Am Math Month 73:231–246
Hwang FK (1982) New concepts in seeding knockout tournaments. Am Math Month 89:235–239
John Molson (2020) John Molson MBA international case competition – Innovate. Connect. Compete. https://mbacasecomp.com/. Accessed 15 Feb 2023
Julian R, Abel R, Finizio NJ, Greig M, Lewis SJ (2003) Generalized whist tournament designs. Discrete Math 268:1–19
Kendall G, Knust S, Ribeiro CC, Urrutia S (2010) Scheduling in sports: an annotated bibliography. Comput Oper Res 37:1–19
Knust S (2021) Classification of literature on sports scheduling. http://www2.inf.uos.de/knust/sportssched/sportlit_class/. Accessed 20 June 2022
Knust S, Lücking D (2009) Minimizing costs in round robin tournaments with place constraints. Comput Oper Res 36:2937–2943

Lamghari A, Ferland JA (2005) Heuristic techniques to assign judges in competitions. In: Prahlad V, Tan WW, Loh AP (eds) Proceedings of the 3rd international conference on computational intelligence, robotics and autonomous systems, Singapore
Lamghari A, Ferland JA (2007) Structured neighborhood tabu search for assigning judges to competitions. In: Srinivasan D (ed) Proceedings of IEEE symposium on computational intelligence in scheduling, pp 238–245
Lamghari A, Ferland JA (2010) Metaheuristic methods based on tabu search for assigning judges to competitions. Ann Oper Res 180:33–61
Lamghari A, Ferland JA (2011) Assigning judges to competitions of several rounds using tabu search. Eur J Oper Res 210:694–705
Li M, Davari M, Goossens D (2022) Multi-league sports scheduling with different leagues sizes. Eur J Oper Res 307: 313–327. https://doi.org/10.1016/j.ejor.2022.10.010
Marchand E (2002) On the comparison between standard and random knockout tournaments. J R Stat Soc D 51:169–178
Melo RA, Urrutia S, Ribeiro CC (2009) The traveling tournament problem with predefined venues. J Schedul 12:607–622
Miyashiro R, Matsui T (2006) Minimizing the carry-over effects value in a round robin tournament. In: Burke EK, Rudová H (eds) Proceedings of the 6th international conference on the practice and theory of automated timetabling, pp 460–463
Oliveira L, Souza CC, Yunes T (2016) Lower bounds for large traveling umpire instances: new valid inequalities and a branch-and-cut algorithm. Comput Oper Res 72:147–159
Ordonez R, Knowles TW (1998) Solving the American league umpire crew scheduling problem as a constraint satisfaction problem. In: Proceedings of the 29th annual meeting of the decision sciences institute, vol 2, pp 1058–1061
Owen P, Weatherston CR (2004) Uncertainty of outcome, player quality and attendence at NPC rugby union matches. J Sports Econ 5:347–370
Rasmussen RV, Trick MA (2006) The timetable constrained distance minimization problem. In: Beck JC, Smith BM (eds) Integration of AI and OR techniques in constraint programming for combinatorial optimization problems. Lecture notes in computer science, vol 3990. Springer, Berlin, pp 167–181
Rasmussen RV, Trick MA (2008) Round robin scheduling – a survey. Eur J Oper Res 188:617–636
Regin JC (2001) Minimization of the number of breaks in sports scheduling problems using constraint programming. In: Freuder E, Wallace R (eds) Constraint programming large scale discrete optimization. DIMACS series in discrete mathematics and theoretical computer science, vol 57, pp 115–130
Ribeiro CC (2012) Sports scheduling: problems and applications. Int Trans Oper Res 19:201–226
Ribeiro CC, Urrutia S (2007) Heuristics for the mirrored traveling tournament problem. Eur J Oper Res 179:775–787
Rokosz FM (2000) Procedures for structuring and scheduling sports tournaments: elimination, consolation, placement, and round-robin design, 3rd edn. Charles C. Thomas, Springfield
Scarf P, Yusof MM, Bilbao M (2009) A numerical study of designs for sporting contests. Eur J Oper Res 198:190–198
Schwenk AJ (2000) What is the correct way to seed a knockout tournament? Am Math Month 107:140–150
Scimia E (2019) What is the Swiss System? https://www.thesprucecrafts.com/the-swiss-system-611537. 20 June 2022
Suksompong W (2016) Scheduling asynchronous round-robin tournaments. Oper Res Lett 44:96–100
Suzuka A, Miyashiro R, Yoshise A, Matsui T (2007) The home-away assignment problems and break minimization/maximization problems in sports scheduling. Pac J Optim 3:113–133
Sziklai BR, Biró P, Csató L (2022) The efficacy of tournament designs. Comput Oper Res 144:105–821

Toffolo TAM, Wauters T, Malderen SV, Berghe GV (2016) Branch-and-bound with decomposition-based lower bounds for the traveling umpire problem. Eur J Oper Res 250:737–744

Trick MA (2004) Using sports scheduling to teach integer programming. INFORMS Trans Edu 5:10–17

Trick MA, Yildiz H, Yunes T (2012) Scheduling major league baseball umpires and the traveling umpire problem. Interfaces 42:232–244

Urrutia S, Ribeiro CC (2006) Maximizing breaks and bounding solutions to the mirrored traveling tournament problem. Discrete Appl Math 154:1932–1938

Van Bulck D, Goossens DR (2022a) Generalizing first-break-then-schedule to time-relaxed sports timetabling. In: Causmaecker PD, Özcan E, Berghe GV (eds) Proceedings of the 13th international conference on the practice and theory of automated timetabling - PATAT 2021: Volume I, pp 172–187

Van Bulck D, Goossens DR (2022b) Optimizing rest times and differences in games played: an iterative two-phase approach. J Schedul 25:262–271

Van Bulck D, Goossens DR, Schönberger J, Guajardo M (2020) RobinX: a three-field classification and unified data format for round-robin sports timetabling. Eur J Oper Res 280:568–580

Van Bulck D, Goossens DR, Beliën J, Davari M (2021) International timetabling competition 2021: sports timetabling – website. https://www.sportscheduling.ugent.be/ITC2021/. 10 Nov 2022

Van Bulck D, Goossens DR, Beliën J, Davari M (2021a) The fifth international timetabling competition (ITC 2021): sports timetabling. In: MathSport international 2021 conference, pp 117–122

Van Bulck D, Goossens DR, Schönberger J, Guajardo M (2021b) RobinX: an XML-driven classification for round-robin sports timetabling - project overview. https://www.sportscheduling.ugent.be/RobinX/. Accessed 10 Nov 2022

de Werra D (1980) Geography, games, and graphs. Discrete Appl Math 2:327–337

de Werra D (1981) Scheduling in sports. In: Hansen P (ed) Studies on graphs and discrete programming. Annals of discrete mathematics, vol 11. North-Holland, Amsterdam, pp 381–395

de Werra D (1988) Some models of graphs for scheduling sports competitions. Discrete Appl. Math 21:47–65

Xue L, Luo Z, Lim A (2015) Two exact algorithms for the traveling umpire problem. Eur J Oper Res 243:932–943

Yavuz M, Inan UH, Figlali A (2008) Fair referee assignments for professional football leagues. Comput Oper Res 35:2937–2951

Yi X, Goossens DR (2023) Strategies for dealing with uncertainty in time-relaxed sports timetabling. Ann Oper Res 320:473–492

Chapter 2
Combinatorial Structures

Mathematical models based on theoretical graph concepts help to formulate and solve fundamental scheduling problems. In this chapter, we consider some elementary sports scheduling problems and concentrate on the combinatorial structures involved in their models and solution approaches.

In practical situations where tournament schedules of sports leagues have to be constructed, the organizers will have various additional requirements to consider. The basic combinatorial models will often need extensions to produce by themselves schedules that may be used in such situations. However, the problem representation using graphs or other combinatorial structures will still be crucial to developing adequate and sometimes intricate procedures that can handle the many constraints in the real scheduling problem.

Overall, this chapter is structured in three parts. The first part encompasses the fundamentals, models, and problems. Graph theory is the main tool used in this chapter, whose first section presents its main elements, definitions, and results relevant to this book. Section 2.2 introduces the basic graph model used to represent single round-robin tournaments and the circle method for constructing timetables. Section 2.3 introduces the definitions of home and away games and the concept of home-away patterns to advance from timetables to schedules. It also defines the concept of breaks in the perfect alternation of home and away games on each team's schedule. Section 2.4 discusses the construction of timetables with specific additional constraints. Section 2.5 extends the definition of single round-robin tournaments to multi-round-robin tournaments, in which each team plays against every other team a fixed number of times. Of particular importance are double round-robin tournaments, a typical design, e.g., of many football leagues.

The second part and the following sections focus on optimal schedules for some specific problem variants. Section 2.6 discusses the structure of schedules with a minimum number of breaks. Section 2.7 introduces two decomposition approaches for constructing schedules. Section 2.8 addresses the minimization of the number of rounds where breaks occur. Fairness is another criterion for scheduling sports

C. C. Ribeiro et al., *Combinatorial Models for Scheduling Sports Tournaments*,
EURO Advanced Tutorials on Operational Research,
https://doi.org/10.1007/978-3-031-37283-4_2

tournaments and Sect. 2.9 considers schedules where all teams should have the same number of breaks. Section 2.10 shows that minimizing the total number of travels is equivalent to maximizing the total number of breaks. We have assumed up to this point that each team has its venue or stadium. Section 2.11 considers balanced tournaments, in which there are shared venues that impose special requirements. Section 2.12 considers a scheduling problem related to assigning referees to games. The minimization of carry-over effects is discussed in Sect. 2.13.

The third part of this chapter discusses more advanced topics. We have shown that constructing a timetable for a sports tournament can be formulated as an edge-coloring problem in a complete graph. Section 2.14 considers an extension of this model, where, for each edge, there is a list of feasible colors. It applies to situations where some games must be played in specific rounds. Section 2.15 discusses edge-recoloring strategies for exploring the landscape of edge colorings. Finally, Sect. 2.16 shows how to find general edge colorings.

2.1 Elements of Graph Theory

This section introduces the main graph theory definitions and results that are essential for understanding the models and algorithms presented in this book. We invite the reader to get acquainted and become familiar with these ideas as they will be used in all chapters.

A graph $G = (V, E)$ consists of a set V of nodes and a set E of edges. Each edge $e = [u, v]$ is an unordered pair of distinct nodes. The nodes $u, v \in V$ are the endpoints of edge e. An edge $e = [u, v]$ may be given an orientation, say from u to v: in this case, it becomes an arc $a = (u, v)$.

A *simple graph* has neither loops (edges whose endpoints coincide) nor parallel edges (multiple edges between the same pair of endpoints). A *multigraph* may have more than one edge connecting the same pair of nodes.

The *degree* $d_G(v)$ of a node $v \in V$ in G is the number of edges having v as an endpoint (i.e., the number of edges adjacent to v). A graph $G = (V, E)$ is *complete* if there is an edge in E between every pair of nodes $u, v \in V$, with $u \neq v$. We denote by K_n the complete graph with $n = |V|$ nodes. The graph G is d-*regular* if $d_G(v) = d$ for every node $v \in V$. A *cycle* is a sequence of edges $[u_1, u_2], [u_2, u_3], \ldots, [u_{p-1}, u_p], [u_p, u_1]$, where $u_1, u_2, \ldots, u_{p-1}, u_p$ are nodes of V.

A graph $G = (V, E)$ is *bipartite* if its node set V may be partitioned into two subsets V_1 and V_2 such that no edge in E has both endpoints in V_1 or in V_2. If every possible edge between V_1 and V_2 exists in the graph, then G is a *complete bipartite* graph. The complete bipartite graph with n nodes in one subset and m nodes in the other is denoted by $K_{n,m}$. A graph is bipartite if and only if it has no *odd cycle* (i.e., a cycle with an odd number of edges).

An edge subset $F \subseteq E$ *spans* the graph $G = (V, E)$ if every node $v \in V$ is the endpoint of at least one edge in F. In this case, F is called a *factor* of G. F is an ℓ-factor of G if every node $v \in V$ is an endpoint of exactly ℓ distinct edges in F.

A subset $F \subseteq E$ such that no two edges in F have a common node is called a *matching*. In a *perfect matching* F (also called a *1-regular matching*), each node $v \in V$ is adjacent to exactly one edge in F. We notice that a perfect matching is equivalent to a 1-factor.

A partition of the edge set E into matchings is an *edge coloring* of graph G. If all matchings in an edge coloring are perfect (i.e., 1-factors), then the edge coloring is a *1-factorization*.

The minimum number of matchings in an edge coloring of graph G is its *chromatic index* $\chi'(G)$. Clearly, $\chi'(G) \geq \Delta(G)$, where $\Delta(G)$ is the largest node degree in G.

Two 1-factorizations $\mathscr{F} = \{F_1, \ldots, F_k\}$ and $\mathscr{H} = \{H_1, \ldots, H_k\}$ of G are called *isomorphic* if there exists a bijective function φ from the node-set V of G onto itself such that $\{F_1^{\varphi}, \ldots, F_k^{\varphi}\} = \{H_1, \ldots, H_k\}$. Here, F_i^{φ} denotes the set of all edges $[\varphi(x), \varphi(y)]$ where $[x, y]$ is an edge in F_i. The concept of isomorphic 1-factorizations can also be extended to *isomorphic edge colorings*.

A 1-factorization is said to be *perfect* when the edges in $F_i \cup F_j$ form a *Hamiltonian cycle* for each pair of distinct 1-factors F_i and F_j.

Remark 2.1 In a complete graph K_{2n}, any 1-factorization is equivalent to a minimum cardinality edge coloring. ■

This remark highlights that 1-factorizations are a particular case of edge colorings. We prefer to use the term 1-factorization. When it is necessary because the coloring does not have the characteristics of a 1-factorization, we use the term edge coloring.

Remark 2.2 Formally, there is no order between the 1-factors of the 1-factorization $\mathscr{F} = \{F_1, \ldots, F_k\}$ (or between the colors of an edge coloring). In the context of this book, we refer to the *ordered 1-factorization* $\mathscr{F} = (F_1, \ldots, F_k)$ as an ordered sequence whenever the order in which the 1-factors (or colors) are taken is relevant. ■

This second remark and the notation defined therein are relevant because, in the context of sports scheduling, the order of the factors of a 1-factorization matters. By analogy, we could also refer to an ordered edge coloring.

2.2 Basic Graph Model for Single Round-Robin Tournaments

In the scheduling problems described in this book, we assume that $2n$ teams (or competitors) form a league that plays one or more tournaments. The case of an odd number of teams may be handled by adding a dummy team to the league.

It is quite natural to associate the $2n$ teams, numbered $1, \ldots, 2n$, with the nodes of a graph. Each game to be played by teams i and j will be represented by an edge $[i, j]$.

There is a one-to-one correspondence between ordered 1-factorizations and single round-robin timetables. If we have to construct a timetable for the games of a single round-robin tournament, we have to find an edge coloring of the edges of a complete graph K_{2n} with $2n$ nodes. Since K_{2n} is $(2n-1)$-regular, any timetable will require at least $2n - 1$ rounds, assuming each team plays at most one game in each round. Indeed, one can always construct a timetable using exactly $2n-1$ rounds (i.e., $\chi'(K_{2n}) = 2n-1$). Let $\mathscr{F} = (F_1, \ldots, F_{2n-1})$ be such an edge coloring, where F_i is the 1-factor whose edges correspond to the games played in round $i = 1, \ldots, 2n-1$. $\mathscr{F}$ is indeed a 1-factorization of K_{2n} since each team has to play exactly one of its $2n - 1$ games during each one of the $2n - 1$ rounds of the timetable.

The *circle method* (or *polygon method*) described in Algorithm 1 provides a simple construction for the so-called *canonical 1-factorization* $\mathscr{F} = (F_1, \ldots, F_{2n-1})$ of K_{2n}, where

$$\mathrm{mod}^{(2n-1)}(x) = \begin{cases} x + 2n - 1, & \text{if } x \leq 0, \\ x, & \text{if } 1 \leq x \leq 2n - 1, \\ x - (2n - 1), & \text{if } x \geq 2n. \end{cases} \tag{2.1}$$

Algorithm 1: Circle method

Input: Set of teams $1, 2, \ldots, 2n$
Output: Canonical 1-factorization $\mathscr{F} = (F_1, \ldots, F_{2n-1})$

for $i = 1$ *to* $2n - 1$ **do**
  $F_i \leftarrow \{[i, 2n]\}$;
  for $\ell = 1$ *to* $n - 1$ **do**
    $F_i \leftarrow F_i \cup \{[\mathrm{mod}^{(2n-1)}(i + \ell), \mathrm{mod}^{(2n-1)}(i - \ell)]\}$;
  end for
end for

We give in Fig. 2.1 a graphical interpretation of the circle method. We first position the teams 1 to $2n - 1$ in the vertices of a regular polygon. For round 1, we first match the highest team of the polygon (team 1) with that in the center (team $2n$). Next, we pair each team on the left side of the polygon with that on the right side at the same height. This perfect matching defines the games of round 1. We

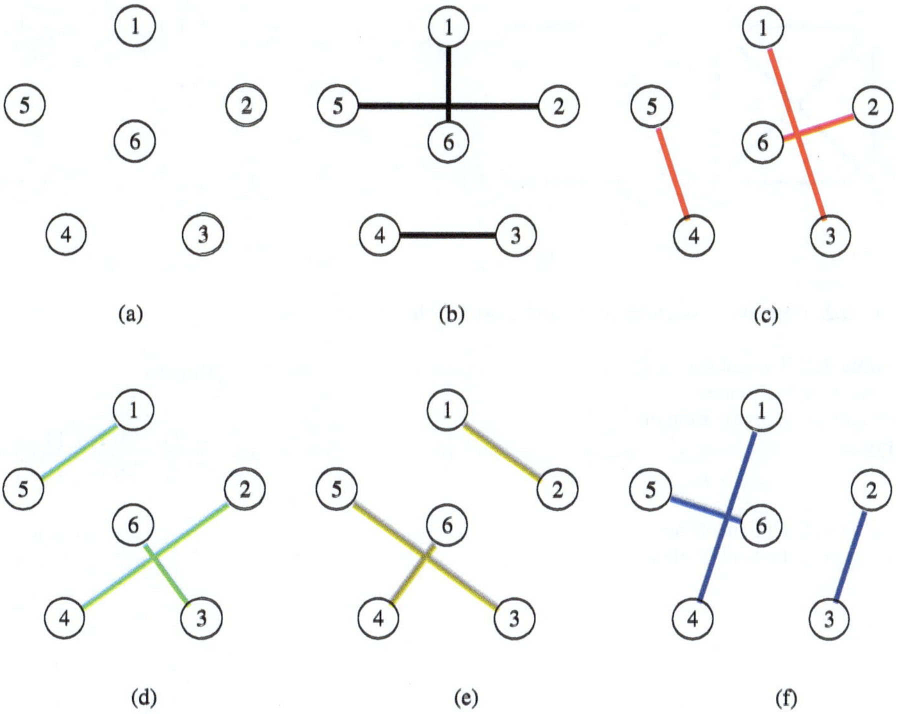

Fig. 2.1 Circle method for a tournament with $2n = 6$ teams. (**a**) Initial setup. (**b**) Round 1. (**c**) Round 2. (**d**) Round 3. (**e**) Round 4. (**f**) Round 5

repeat this procedure for rounds $2, 3, \ldots, 2n - 1$, as if the teams in the polygon's vertices were rotated counterclockwise at each step (or round).

The chronological sequence of the rounds in which the games are scheduled in the 1-factorization $\mathscr{F} = (F_1, \ldots, F_{2n-1})$ induces an order on $F_1, \ldots, F_{2n-1}$. However, we shall refer to 1-factorizations without mentioning the order whenever no confusion may arise.

The circle method is used in some of the problems and models described later in this book. Figure 2.2 shows a canonical 1-factorization $\mathscr{F} = (F_1, F_2, F_3)$ of K_4. In the first round, team 4 plays a game with team 1, while team 2 plays against team 3, then $F_1 = \{[4, 1], [2, 3]\}$. Similarly, $F_2 = \{[4, 2], [3, 1]\}$ and $F_3 = \{[4, 3], [1, 2]\}$.

Table 2.1 represents the timetable corresponding to the canonical 1-factorization of Fig. 2.2.

An *array* is a data structure associated with a systematic arrangement of similar objects, usually in rows and columns. One may also represent the timetable of a single round-robin tournament of a league of $2n$ teams by a 2-dimensional opponents array with $2n$ rows and $2n - 1$ columns. Each row corresponds to a team, while each column is associated with a round of the timetable. The entry in row i and column r of this array gives the opponent of team i in round r. For example,

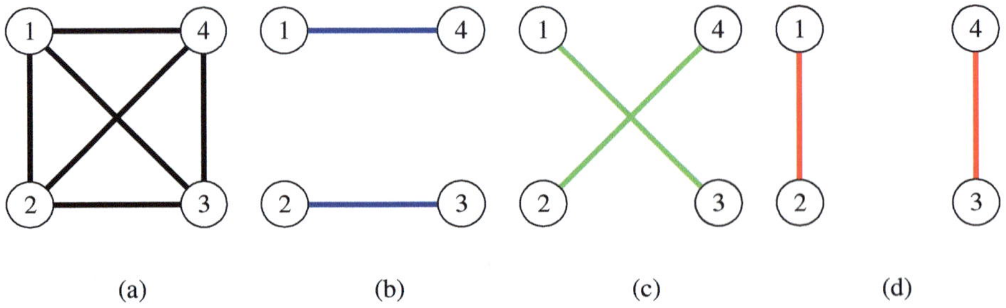

Fig. 2.2 Canonical 1-factorization of K_4. **(a)** K_4. **(b)** F_1. **(c)** F_2. **(d)** F_3

Table 2.1 Timetable corresponding to the canonical 1-factorization of Fig. 2.2

Rounds		
1	2	3
[4, 1]	[4, 2]	[4, 3]
[2, 3]	[3, 1]	[1, 2]

Table 2.2 Opponents array for the timetable of Table 2.1

	Rounds		
Teams	1	2	3
1	4	3	2
2	3	4	1
3	2	1	4
4	1	2	3

the entry in row 2 and column 3 in Table 2.2 denotes that team 2 meets team 1 in round 3.

Remark 2.3 In leagues with an odd number of teams, there is (at least) one team that does not play a game in each round. We say that such a team has a *bye* in this round. For a single round-robin tournament represented by the complete graph K_{2n-1}, each team must play $2n-2$ games. Therefore, any timetable will require at least $2n-1$ rounds. One has $\chi'(K_{2n-1}) = 2n-1$, and one can construct a timetable with $2n-1$ rounds by starting from a timetable $\mathscr{F}$ of K_{2n} and remove all games of team $2n$ (which does not exist and may be seen as a dummy team in a league of $2n$ teams). ■

There are many more 1-factorizations for complete graphs besides the canonical ones. There is only one non-isomorphic 1-factorization for K_2, K_4, and K_6, but this number increases very fast: there are six for K_8, 396 for K_{10}, and 526,915,620 for K_{12}. We often need to use these non-canonical 1-factorizations, particularly when the problem constraints will restrict the set of rounds in which some specific games may be scheduled. We show in Sect. 2.16 another algorithm that can construct any 1-factorization of K_{2n}, including the canonical.

2.3 From Timetables to Schedules: Home and Away Games

It is often the case that each team has its venue (or a stadium), and each game between teams i and j is played either in the home city of team i or the home city of team j. In the first case, we say that team i plays a *home game*, and that team j plays an *away game*. It is convenient to represent this game by an arc (j, i) oriented from j to i. In the second case, if the game between teams i and j is played in the home city of team j, we have an arc (i, j) oriented from i to j.

If every game in a single round-robin tournament is assigned to a venue, then each edge of K_{2n} receives an orientation and becomes an arc. Instead of an edge coloring, we now have an *oriented edge coloring*, denoted by $\vec{\mathscr{F}} = (\vec{F}_1, \ldots, \vec{F}_{2n-1})$.

For the timetable associated to K_4 in Tables 2.1 and 2.2, the edges may be oriented, e.g., as

$$\vec{F}_1 = \{(4, 1), (3, 2)\}, \vec{F}_2 = \{(2, 4), (3, 1)\}, \text{ and } \vec{F}_3 = \{(4, 3), (1, 2)\},$$

as represented in Fig. 2.3.

We can associate a *home-away pattern* (or a HAP) with the oriented edge coloring $\vec{\mathscr{F}} = (\vec{F}_1, \cdots, \vec{F}_{2n-1})$ and represent it by an array named *HAP* with $2n$ rows and $2n - 1$ columns, where $HAP(i, r)$ = "H" if team i plays a home game in round r, $HAP(i, r)$ = "A" if team i plays an away game in round r, or $HAP(i, r)$ = " $*$ " if team i has a bye in round r, for $i = 1, \ldots, 2n$ and $r = 1, \ldots, 2n - 1$. Together, the timetable and the home-away pattern define the tournament *schedule*.

Row i of the HAP gives the *profile* of team i. Usually, one tries to alternate as much as possible the H's and the A's in the profile of each team, i.e., the home and away games played by this team. No two teams may have the same profile. To show it, we suppose that two teams had the same profiles. Since both teams would be simultaneously away or both at home, they would never be able to play against each

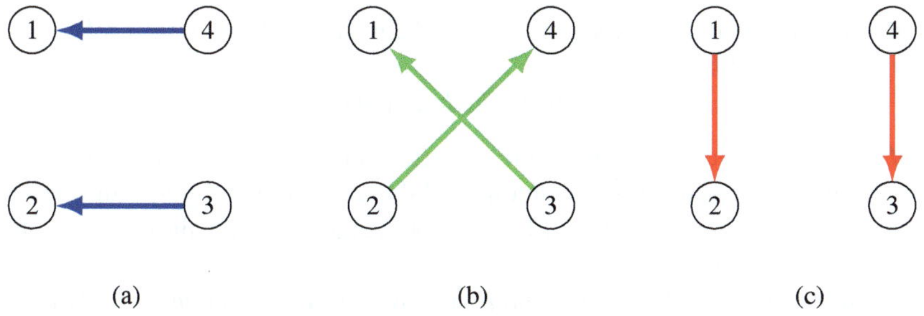

Fig. 2.3 Oriented edge coloring for the edge coloring of Fig. 2.2. **(a)** $\vec{F}_1$. **(b)** $\vec{F}_2$. **(c)** $\vec{F}_3$

Table 2.3 Home-away pattern (HAP) associated with the oriented edge coloring of Fig. 2.3 (breaks are underlined)

	Rounds		
Teams	1	2	3
1	H	<u>H</u>	A
2	H	A	H
3	A	<u>A</u>	H
4	A	H	A

other. Therefore, at most two teams may have perfectly alternating profiles: one that starts the tournament with a home game and another that starts with an away game.

If the profile of a team i has two consecutive entries in rounds $r-1$ and r with the same symbol (H or A), we say that team i has a *break* (of alternation) in round r. The HAP in Table 2.3 illustrates that teams 1 and 3 have a break in round 2. If a team plays two consecutive home games, we say it has a *home break* (as for team 1 in the table). If a team plays two consecutive away games, we say it has an *away break* (as for team 3 in the table).

We now recall some elementary properties of schedules.

Proposition 2.1 *Any schedule of a single round-robin tournament with* $2n$ *teams has at least* $2n-2$ *breaks.*

Proof If an oriented edge coloring exists with $p < 2n-2$ breaks, there are at least $2n - p \geq 3$ teams without breaks in their profiles. Therefore, two teams, i and j have the same profile. It follows that i and j can never play against each other. ∎

This result shows that the oriented edge coloring of K_4 in Fig. 2.3 has a minimum number $2n-2=2$ of breaks.

Remark 2.4 One can also consider oriented edge colorings of graphs that are not complete and similarly define breaks by simply ignoring byes. There is a break (home or away) for team i in round r if entries $HAP(i, r')$ and $HAP(i, r)$, with $r' < r$, of the home-away pattern are equal and all entries $HAP(i, r'+1), \ldots, HAP(i, r-1)$ are byes. ∎

Proposition 2.2

(i) *For a single round-robin tournament with* $2n$ *teams, there exists a schedule with exactly* $2n-2$ *breaks.*
(ii) *For* $2n-1$ *teams, there exists a schedule without breaks.*

Proof In what follows, we give the construction of a canonical oriented edge coloring of K_{2n}. It starts from a canonical 1-factorization and gives an orientation to its edges so that teams 1 and $2n$ have no breaks and the remaining $2n-2$ teams have one break each.

The method is called the *canonical procedure for edge orientation* and its template appears in Algorithm 2. Given a canonical 1-factorization $\mathscr{F} = (F_1, \ldots, F_{2n-1})$ obtained as described in Sect. 2.2, the procedure gives an orientation $\overrightarrow{F}_i$ to each 1-factor F_i, for $i = 1, \ldots, 2n-1$, creating an oriented canonical 1-factorization $\vec{\mathcal{F}} = (\vec{F}_1, \ldots, \vec{F}_{2n-1})$. Table 2.4 illustrates an example for $2n = 6$.

Algorithm 2: Canonical procedure for edge orientation

Input: Canonical 1-factorization $\mathcal{F} = (F_1, \ldots, F_{2n-1})$
Output: Oriented canonical 1-factorization $\vec{\mathcal{F}} = (\vec{F}_1, \ldots, \vec{F}_{2n-1})$

1 **for** $i = 1$ *to* $2n - 1$ **do**
2   **if** i *is even* **then**
3   Edge $[i, 2n]$ becomes arc $(2n, i)$;
4   **else**
5   Edge $[i, 2n]$ becomes arc $(i, 2n)$;
6   **end if**
7   **for** $\ell = 1$ *to* $n - 1$ **do**
8   **if** ℓ *is odd* **then**
9   Edge $[\mathrm{mod}^{(2n-1)}(i + \ell), \mathrm{mod}^{(2n-1)}(i - \ell)]$ becomes arc $(\mathrm{mod}^{(2n-1)}(i + \ell), \mathrm{mod}^{(2n-1)}(i - \ell))$;
10   **else**
11   Edge $[\mathrm{mod}^{(2n-1)}(i + \ell), \mathrm{mod}^{(2n-1)}(i - \ell)]$ becomes arc $(\mathrm{mod}^{(2n-1)}(i - \ell), \mathrm{mod}^{(2n-1)}(i + \ell))$;
12   **end if**
13   **end for**
14 **end for**

We observe that team i has a unique break for any $2 \leq i \leq 2n-1$. It occurs either in the round of game $(i, 2n)$ (if i is odd) or in the round following game $(2n, i)$ (if i is even). Teams 1 and $2n$ have no breaks. Therefore, for obtaining an oriented edge coloring of K_{2n-1}, we omit all games involving team $2n$, which gives a schedule in $2n - 1$ rounds without breaks. ■

Table 2.4 Canonical oriented 1-factorization of K_6. The arcs in red are oriented from node i to $2n$ (line 5) and from node $i - \ell$ to $i + \ell$ (line 11), while the arcs in blue are oriented from node $2n$ to i (line 3) and from node $i + \ell$ to $i - \ell$ (line 9) in Algorithm 2

Canonical 1-factorization	Canonical oriented 1-factorization
$F_1 = \{[6, 1], [2, 5], [3, 4]\}$	$\vec{F}_1 = \{(1, 6), (2, 5), (4, 3)\}$
$F_2 = \{[6, 2], [3, 1], [4, 5]\}$	$\vec{F}_2 = \{(6, 2), (3, 1), (5, 4)\}$
$F_3 = \{[6, 3], [4, 2], [5, 1]\}$	$\vec{F}_3 = \{(3, 6), (4, 2), (1, 5)\}$
$F_4 = \{[6, 4], [5, 3], [1, 2]\}$	$\vec{F}_4 = \{(6, 4), (5, 3), (2, 1)\}$
$F_5 = \{[6, 5], [1, 4], [2, 3]\}$	$\vec{F}_5 = \{(5, 6), (1, 4), (3, 2)\}$

2.4 Another Type of 1-Factorization

Sometimes local constraints may prevent the construction of a timetable based on canonical 1-factorizations.

In the case described here, the timetable of a single round-robin tournament for a league of $2n$ teams corresponds to a 1-factorization $F_1, F_2, \ldots, F_{2n-1}$ of K_{2n}, where the 1-factors $F_1, F_3, \ldots, F_{2n-3}, F_{2n-1}$ are weekend games, while $F_2, F_4, \ldots, F_{2n-2}$ are 1-factors associated with weekday games.

The $2n$ teams are partitioned into two subsets X and Y of n teams each (we assume that n is even), considering the distances between them. The distance between any team in set X and any team in set Y should be greater than the distance between any two teams within the same set.

To avoid long trips during the weekdays, we force any 1-factor F_i for which i is odd (i.e., $i = 1, 3, \ldots, 2n - 1$) to have only games pairing a team of X with a team of Y, while any 1-factor F_i for which i is even (i.e., $i = 2, 4, \ldots, 2n$) consists of games pairing exclusively two teams of the same subset. This 1-factorization can easily be constructed.

Since n is even, there are 1-factorizations of the complete graphs K_n^X and K_n^Y, associated with X and to Y, respectively. They can be combined to produce the 1-factors $F_2, F_4, \ldots, F_{2n-2}$.

The games pairing a team in X with a team in Y form a complete bipartite graph $K_{n,n}$ with a bipartition $\{X, Y\}$, with n nodes on each side (recall that n is even). This graph has a 1-factorization that gives $F_1, F_3, \ldots, F_{2n-1}$.

Since the game between two teams i and j has to be played at the venue of either one of them, we have to assign the status of home or away to all teams in all rounds. There is a construction that gives a schedule with a minimum number of breaks, namely $2n - 2$:

Proposition 2.3 *Consider a single round-robin tournament of a league with $2n$ teams divided into two equal-sized subsets X and Y, such that $|X| = |Y|$ is even. One may construct an oriented 1-factorization $\vec{F}_1, \vec{F}_2, \ldots, \vec{F}_{2n-1}$ of K_{2n} such that:*

- *the oriented 1-factors $\vec{F}_1, \vec{F}_3, \ldots, \vec{F}_{2n-1}$ of K_{2n} consist of arcs with one endpoint in X and the other in Y;*
- *the oriented 1-factors $\vec{F}_2, \vec{F}_4, \ldots, \vec{F}_{2n-2}$ of K_{2n} consist of arcs with both endpoints in the same subset X or Y; and*
- *there are $2n - 2$ breaks.* ■

The construction is illustrated in Fig. 2.4 for a single round-robin tournament of $2n = 8$ teams. The nodes in gray represent teams with a break in the corresponding weekend round. Considering $X = \{x_1, x_2, x_3, x_4\}$ and $Y = \{y_1, y_2, y_3, y_4\}$, then teams y_1 and y_3 have a break in round 3, x_1 and x_3 have a break in round 5, y_2 and y_4 have a break in round 7, and x_2 and x_4 have no break.

We note that if n was odd, this construction would not be possible because each 1-factor F_i (with i even) would have to contain also an edge between X and Y.

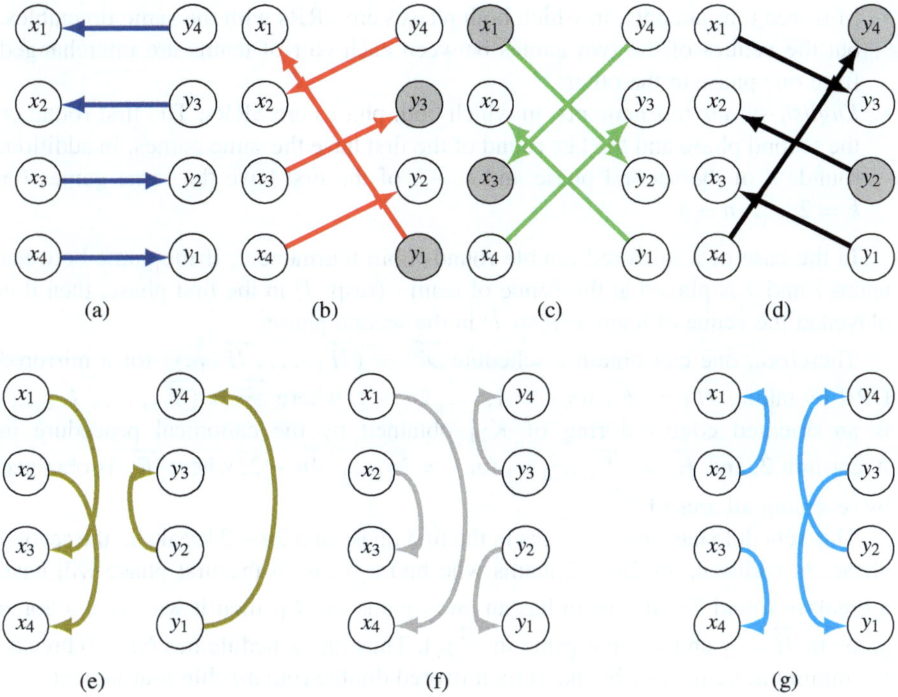

Fig. 2.4 Oriented 1-factorization for a single round-robin tournament with $2n = 8$ teams divided into two subsets X and Y of four teams (nodes in gray represent teams with a break in the corresponding weekend round). **(a)** $\overrightarrow{F}_1$. **(b)** $\overrightarrow{F}_3$. **(c)** $\overrightarrow{F}_5$. **(d)** $\overrightarrow{F}_7$. **(e)** $\overrightarrow{F}_2$. **(f)** $\overrightarrow{F}_4$. **(g)** $\overrightarrow{F}_6$

2.5 Multi-Round-Robin Tournaments

Most sports leagues play either a single round-robin tournament (SRR), where each pair of teams meets once, or a double round-robin tournament (DRR), where each pair of teams meets twice. In the general case, each team plays against every other team exactly m times. The number of rounds needed to schedule all $m\binom{2n}{2}$ games is equal to $m(n-1)$, where each team plays exactly one game in each round. Double-round-robin tournaments can be organized in different manners, as mentioned in Sect. 1.1.3.1:

- Unconstrained DRR tournaments, in which the order of the games is arbitrary.
- Tournaments with the *no-repeaters* constraint, which does not allow the two games between the same pair of teams to be played in two consecutive rounds.
- Tournaments in which a minimum number of rounds should separate the two games between the same pair of teams.
- Tournaments organized in two separate phases, each of which is an SRR tournament, possibly with a different timetable.

- Mirrored tournaments, in which both phases are SRRs with the same timetables, but the venues of the two games between each pair of teams are interchanged from one phase to the other.
- *English-system* tournaments, in which both phases are SRRs. The first round of the second phase and the last round of the first have the same games. In addition, rounds k of the second phase and $k-1$ of the first have the same games for $k = 2, \ldots, n-1$.

In the case of a mirrored double round-robin tournament, if the game between teams i and j is played at the venue of team i (resp. j) in the first phase, then it is played at the venue of team j (resp. i) in the second phase.

Therefore, one can obtain a schedule $\vec{\mathscr{H}} = (\vec{H}_1, \ldots, \vec{H}_{4n-2})$ for a mirrored DRR by taking $\vec{H}_i = \vec{F}_i$ for $i = 1, \ldots, 2n-1$, where $\vec{\mathscr{F}} = (\vec{F}_1, \ldots, \vec{F}_{2n-1})$ is an oriented edge coloring of K_{2n} obtained by the canonical procedure in Algorithm 2, and $\vec{H}_i = \overleftarrow{F}_{i-(2n-1)}$ for $i = 2n, \ldots, 4n-2$, where $\overleftarrow{F}_i$ is obtained by reversing all arcs of $\vec{F}_i$.

This schedule has $2n-2$ breaks in the first phase and $2n-2$ breaks in the second phase. In addition, all $2n-2$ teams who had a break in the first phase will have a break in round $2n$ (if a team has an away game in $\vec{H}_1$, then it will have a home game in $\vec{H}_{2n-1}$ and a home game in $\vec{H}_{2n}$). Thus, the schedule has $6n-6$ breaks, the minimum number of breaks for a mirrored double round-robin tournament.

This construction may lead to teams with two consecutive breaks (i.e., three consecutive home games or three consecutive away games), which might be undesirable in some situations. Such a situation may occur in round $2n$ of a double round-robin tournament. The teams with a break in round $2n-1$ are teams $2n-2$ and $2n-1$. We may move these breaks to earlier rounds so that no team has a break in round $2n-1$ by modifying line 2 of the original canonical procedure for edge orientation, as it appears now in Algorithm 3.

The modified oriented edge coloring of K_6 appears in Fig. 2.5. We notice that these modifications induce the following changes in the HAP:

- Team $2n-3$ had a break in round $2n-3$, which moved to round $2n-2$.
- Team $2n-2$ had a break in round $2n-1$, which moved to round $2n-2$.
- Team $2n-1$ had a break in round $2n-1$ that disappeared.
- Team $2n$ had no break before but acquired a break in round $2n-3$.

Since there are now no breaks in round $2n-1$, it follows that for a double round-robin tournament, there is a schedule with $6n-6$ breaks such that no team has two consecutive breaks.

Algorithm 3: Modified canonical procedure for edge orientation for mirrored DRR tournaments

Input: Canonical 1-factorization $\mathcal{F} = (F_1, \ldots, F_{2n-1})$
Output: Oriented canonical 1-factorization $\vec{\mathcal{F}} = (\vec{F}_1, \ldots, \vec{F}_{2n-1})$

for $i = 1$ *to* $2n - 1$ **do**
  if *(i is odd and* $i \leq 2n - 5$*) or* $i = 2n - 2$ **then**
    Edge $[i, 2n]$ becomes arc $(i, 2n)$;
  else
    Edge $[i, 2n]$ becomes arc $(2n, i)$;
  end if
  for $\ell = 1$ *to* $n - 1$ **do**
    if ℓ *is odd* **then**
      Edge $[\text{mod}^{(2n-1)}(i + \ell), \text{mod}^{(2n-1)}(i - \ell)]$ becomes arc $(\text{mod}^{(2n-1)}(i + \ell), \text{mod}^{(2n-1)}(i - \ell))$;
    else
      Edge $[\text{mod}^{(2n-1)}(i + \ell), \text{mod}^{(2n-1)}(i - \ell)]$ becomes arc $(\text{mod}^{(2n-1)}(i - \ell), \text{mod}^{(2n-1)}(i + \ell))$;
    end if
  end for
end for

Oriented canonical 1-factorization
$\vec{F}_1 = \{(1,6),(2,5),(4,3)\}$
$\vec{F}_2 = \{(6,2),(3,1),(5,4)\}$
$\vec{F}_3 = \{(6,3),(4,2),(1,5)\}$
$\vec{F}_4 = \{(4,6),(5,3),(2,1)\}$
$\vec{F}_5 = \{(6,5),(1,4),(3,2)\}$

	Rounds				
Teams	1	2	3	4	5
1	A	H	A	H	A
2	A	H	H	A	H
3	H	A	H	H	A
4	A	H	A	A	H
5	H	A	H	A	H
6	H	A	A	H	A

Fig. 2.5 Modified oriented 1-factorization of K_6 and its home-away pattern (modified arc orientations appear in red, breaks are underlined)

2.6 On the Structure of HAPs with a Minimum Number of Breaks

We showed in Propositions 2.1 and 2.2 that the minimum number of breaks for a single round-robin schedule with $2n$ teams is $2n - 2$. We now characterize the HAPs associated with schedules with $2n - 2$ breaks.

Proposition 2.4 *The number of home breaks equals the number of away breaks in any round of a round-robin schedule.*

Proof There are n teams playing an away game and n teams playing a home game in each round. If, in any round, there are more home breaks than away breaks, then more teams would play home than away (and vice-versa). ■

Proposition 2.5 *No team has more than one break in a round-robin schedule with* $2n - 2$ *breaks.*

Proof If a team had more than one break, there would be at least three teams without breaks, which is impossible in a round-robin tournament. ■

Proposition 2.6 *There are either 0 or 2 breaks in each round of any round-robin schedule with* $2n - 2$ *breaks.*

Proof Suppose there are $p \geq 4$ breaks in some round r and, without loss of generality, suppose they occur for teams $1, \ldots, p$. From Proposition 2.5, each of these p teams has a unique break, which occurs in round r. However, there are only two profiles with exactly one break in round r, one starting with a home game and the other with an away game. Since all teams must have different profiles, this situation is impossible. ■

We say that two team profiles are *complementary* if one is obtained from the other by interchanging their home and away games. We state the following corollary from the above propositions:

Corollary 2.1 *The HAP associated with a single round-robin tournament schedule of* $2n$ *teams with a minimum number of breaks consists of* n *pairs of complementary profiles. One pair has no break, and the* $n-1$ *other pairs have one break each. Each pair of complementary profiles has its breaks in a different round.* ■

The HAP illustrated in Table 2.3 corresponds to a single round-robin tournament schedule of a league of four teams, where teams 2 and 4 have complementary profiles without any break. Teams 1 and 3 have complementary profiles with one break each in round 2. For the HAP in Fig. 2.5 with six teams, the pairs of teams with complementary profiles are 1–5, 2–6, and 3–4. The first pair of teams has no breaks, while the other pairs have one break each.

For a single round-robin tournament schedule of a league with $2n$ teams, an optimal HAP is entirely determined by providing the $n - 1$ profiles of one team in each of the pairs of complementary profiles with one break each. Alternatively, we may provide the $n - 1$ rounds where the breaks should occur (in pairs).

For example, consider a league with $2n = 6$ teams for which we want to construct a HAP with $2n - 2 = 4$ breaks occurring in pairs in rounds 2 and 5. Figure 2.6a shows three profiles, all starting with an away game. The first profile has no break. The second and third profiles have a break in rounds 2 and 5, respectively. The complete HAP in Fig. 2.6b is obtained by complementing profiles 1, 2, and 3 to obtain profiles 4, 5, and 6.

	Rounds				
Profiles	1	2	3	4	5
1	A	H	A	H	A
2	A	A	H	A	H
3	A	H	A	H	H

(a)

	Rounds				
Profiles	1	2	3	4	5
1	A	H	A	H	A
2	A	A	H	A	H
3	A	H	A	H	H
4	H	A	H	A	H
5	H	H	A	H	A
6	H	A	H	A	A

(b)

Fig. 2.6 Construction of an optimal HAP for $2n = 6$ teams with $2n - 2 = 4$ breaks (breaks are underlined). (**a**) Three pairs of complementary profiles for a league with $2n = 6$ teams. (**b**) Complete HAP with all $2n - 2 = 4$ breaks indicated by underlined entries

2.7 Constructing Schedules from HAPs or from Timetables

In some applications, we may have a given fixed HAP, and the question is to determine whether there is a compatible timetable assigning each game of the single round-robin tournament to the rounds $r = 1, \ldots, 2n - 1$.

Such a timetable may only sometimes exist: consider, for instance, the teams represented by profiles 1, 3, and 5 in the example of Fig. 2.6. Teams 1 and 5 have to meet in the first round since both play home or away in every other round. Similarly, the game between teams 1 and 3 has to be scheduled in round 5. The game between teams 3 and 5 could be scheduled in either round 1 or 5. However, this can not be done since, in both rounds, either team 3 or team 5 would already be assigned to play with team 1.

According to Corollary 2.1, HAPS with the minimum number of breaks have precisely two breaks in $n - 1$ of the $2n - 2$ rounds where they may occur. The example above illustrates that among the $\binom{2n-2}{n-1}$ possible HAPs with the minimum number of beaks of a single round-robin tournament of a league with $2n$ teams, there are some for which no timetable can be constructed.

Constructing a timetable from a given HAP is called a *first-break, then-schedule* decomposition approach; see also Sect. 4.1.

An integer programming formulation can be used to determine if a given HAP (with the minimum number of breaks or not) corresponds to a timetable. Let $h_{ir} = 1$ if team i plays at home in round r according to the HAP; 0, otherwise.

We define the following binary decision variables for every round $r = 1, \ldots, 2n - 1$ and every pair of teams $i, j = 1, \ldots, 2n$, with $i < j$:

$$x_{ijr} = \begin{cases} 1, & \text{if there is a game between teams } i \text{ and } j \text{ in round } r, \\ 0, & \text{otherwise.} \end{cases}$$

The following integer programming problem determines if the HAP is feasible:

$$\max z(x) = \sum_{i=1}^{2n-1} \sum_{j=i+1}^{2n} \sum_{r=1}^{2n-1} x_{ijr}, \tag{2.2}$$

subject to:

$$\sum_{r=1}^{2n-1} x_{ijr} \leq 1, \qquad i = 1, \ldots, 2n-1, \quad j = i+1, \ldots, 2n, \tag{2.3}$$

$$\sum_{j=1}^{i-1} x_{jir} + \sum_{j=i+1}^{2n} x_{ijr} \leq 1, \qquad i = 1, \ldots, 2n, \quad r = 1, \ldots, 2n-1, \tag{2.4}$$

$$x_{ijr} \leq |h_{ir} - h_{jr}|, \quad i = 1, \ldots, 2n-1, \quad j = i+1, \ldots, 2n, \quad r = 1, \ldots, 2n-1, \tag{2.5}$$

$$x_{ijr} \in \{0, 1\}, \qquad i = 1, \ldots, 2n-1, \quad j = i+1, \ldots, 2n, \quad r = 1, \ldots, 2n-1. \tag{2.6}$$

The objective function $z(x)$ to be maximized in (2.2) is the number of games of the given HAP that can be scheduled. A timetable will be obtained if the maximum of $z(x)$ is equal to $n(2n-1)$, i.e., there are n games in each of the $2n-1$ rounds. Constraints (2.3) express that each pair of teams meets at most once. Constraints (2.4) impose that team i has at most one game in each round r. Constraints (2.5) enforce that two teams i and j are allowed to meet in round r if one of the teams plays a home game and the other an away game in this round. Finally, constraints (2.6) are the integrality requirements.

We now discuss the opposite situation, corresponding to a *first-schedule, then-break* decomposition approach: a timetable is given, and one has to find a HAP with a minimum number of breaks; see also Sects. 4.1 and 5.2.4.

Given a timetable for a single round-robin tournament of a league with $2n$ teams, one can find in $O(n^2)$ time whether there exists a HAP with $2n$ breaks associated with this timetable. The algorithm relies on the special structure of HAPs with a minimum number of breaks. More generally, given a timetable, it is difficult to find an associated HAP with a minimum number of breaks, i.e., with $2n-2$ breaks. Furthermore, for a given timetable, the worst case for the minimum number of breaks is $n(n-1)$ for n even or $(n-1)^2$ for n odd.

2.8 Minimizing the Number of Rounds with Breaks

We have seen that HAPs for single round-robin tournaments with a minimum number of breaks have exactly $n - 1$ rounds with breaks. In this section, we are interested in minimizing the number of rounds where breaks occur, which may increase the total number of breaks.

We recall that a graph is bipartite if its node-set may be partitioned into two subsets V_1 and V_2 such that none of its edges has both endpoints in V_1 or in V_2. A bipartite graph has no odd cycle.

Proposition 2.7 *Let $\mathcal{F} = (F_1, \ldots, F_{2n-1})$ be a 1-factorization of K_{2n} that defines a timetable for a single round-robin tournament of a league with $2n$ teams. Then, a schedule without breaks in rounds $r+1, \ldots, r+t$ can be constructed by assigning orientations to the edges of $\mathcal{F}$ if and only if $F_r \cup \ldots \cup F_{r+t}$ is a bipartite $(t+1)$-factor of K_{2n}.*

Proof Let $\mathcal{F}$ be such a 1-factorization and assume that, for some values of r and t, $F_r \cup \ldots \cup F_{r+t}$ is a bipartite $(t+1)$-factor. Let V_1 and V_2 define a partition of the node-set such that all edges in $F_r \cup \ldots \cup F_{r+t}$ have an endpoint in V_1 and the other endpoint in V_2. For every $i = r, \ldots, r+t$, we give the orientation from the node in V_1 to that in V_2 to all edges in F_i, if i is even. Otherwise, we give the orientation from the node in V_2 to that in V_1 to all edges in F_i. By giving an arbitrary orientation to the remaining edges in $\mathcal{F}$, we obtain a schedule where no breaks occur in rounds $r+1, \ldots, r+t$.

Conversely, we assume that $F_r \cup \ldots \cup F_{r+t}$ is not bipartite. Therefore, it contains an odd cycle C. In order to have no breaks between rounds $r+1$ and $r+t$, all teams must have a perfectly alternating profile from round r to round $r+t$. There are only two such profiles, one starting with a home game and the other with an away game. Two adjacent nodes in C (i.e., two teams that have to meet between rounds r and $r+t$) will have the same profile and hence will not be able to meet, which is a contradiction. Therefore, $F_r \cup \ldots \cup F_{r+t}$ is bipartite. ∎

Hence, if the edge-set of K_{2n} is partitioned into p subsets $M_1, \ldots, M_p$, where each M_i is a bipartite t_i-factor (with $t_1 + \cdots + t_p = 2n - 1$), one may construct a schedule by giving adequate orientations to the edges. This construction is done in such a way that breaks occur only in rounds $t_1 + 1, t_1 + t_2 + 1, \ldots, t_1 + t_2 + \ldots + t_{p-1} + 1$. If we minimize p, we also minimize the number of rounds with breaks. For a league with $2n$ teams, the smallest value of p is $\lceil \log_2 2n \rceil$. In fact, one can show that the partition can be constructed in such a way that the values of t_i are either $\lceil (2n-1)/p \rceil$ or $\lfloor (2n-1)/p \rfloor$.

Proposition 2.8 *There is a decomposition of the edge set of K_{2n} into $p = \lceil \log_2 2n \rceil$ bipartite t_i-factors such that $-1 \le t_i - t_j \le +1$ for any $i, j \le p$.* ∎

Figure 2.7 shows a decomposition for K_8 into $\lceil \log_2 8 \rceil = 3$ bipartite factors: the first of them is a 3-factor, and the other two are 2-factors.

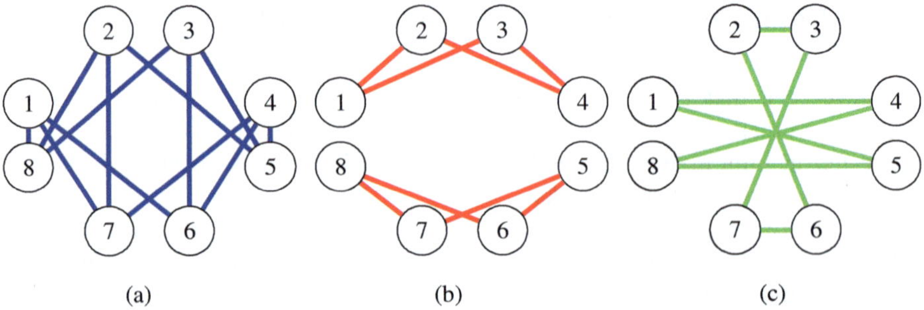

Fig. 2.7 Decomposition of K_8 into a minimum number of bipartite graphs. (**a**) 3-factor. (**b**) First 2-factor. (**c**) Second 2-factor

We can also show that such a schedule with at most $2\lfloor \frac{n}{2} \rfloor \lceil \log_2 n \rceil$ breaks can be found.

2.9 Fairness of Schedules

Fairness is another criterion for tournament scheduling. The first idea is to require that all teams have the same number of breaks. Single round-robin schedules in which every one of the $2n$ teams has one break are not more difficult to obtain than schedules with $2n-2$ breaks. Indeed, from any schedule in which each team has one break, one can derive a schedule with $2n-2$ breaks by a simple cyclic permutation of the rounds (if two teams have a break in round r, then taking the rounds in the order $r, r+1, \cdots, 2n-1, 1, 2, \cdots, r-1$ will give a schedule with $2n-2$ breaks).

Similarly, from any schedule with $2n-2$ breaks, one can get a schedule with $2n$ breaks by adequately permuting the rounds cyclically.

In these schedules, one-half of the teams have an away break, while the other half have a home break. This situation sometimes is considered unfair, and for a schedule to be fair, each team should have the same number of away and home breaks.

When it is required that for each team, the number of home and away breaks must be the same, the best schedule we can have is one in which each team has exactly one home break and one away break. A HAP of this type is represented in Table 2.5 for $2n = 8$ teams.

There is still some unfairness in such a schedule, as we will see. Assume that some team has a larger interval between its two breaks than others. If the team considered has an away break first, this team will likely play many games with low vitality caused by the preceding away break (for two consecutive rounds, the team has not played at home). Conversely, if a team has a home break first, this team may take advantage of its first home break for a longer period.

To prevent this kind of unfairness, we look for schedules where the two breaks of each team are adjacent so that the respective advantages or disadvantages are

Table 2.5 HAP for $2n = 8$ teams where each team has one away and one home break (breaks are underlined)

Teams	Rounds 1	2	3	4	5	6	7
1	A	$\underline{\text{A}}$	H	A	H	$\underline{\text{H}}$	A
2	A	H	$\underline{\text{H}}$	A	$\underline{\text{A}}$	H	A
3	A	H	A	$\underline{\text{A}}$	H	$\underline{\text{H}}$	A
4	A	H	A	H	$\underline{\text{H}}$	A	$\underline{\text{A}}$
5	H	$\underline{\text{H}}$	A	H	A	$\underline{\text{A}}$	H
6	H	A	$\underline{\text{A}}$	H	$\underline{\text{H}}$	A	H
7	H	A	H	$\underline{\text{H}}$	A	$\underline{\text{A}}$	H
8	H	A	H	A	$\underline{\text{A}}$	H	$\underline{\text{H}}$

Table 2.6 HAP for $2n = 8$ teams where the two breaks of each team are consecutive (breaks are underlined)

Teams	Rounds 1	2	3	4	5	6	7
1	A	$\underline{\text{A}}$	H	$\underline{\text{H}}$	A	H	A
2	A	H	$\underline{\text{H}}$	A	$\underline{\text{A}}$	H	A
3	A	H	A	$\underline{\text{A}}$	H	$\underline{\text{H}}$	A
4	A	H	A	H	$\underline{\text{H}}$	A	$\underline{\text{A}}$
5	H	$\underline{\text{H}}$	A	$\underline{\text{A}}$	H	A	H
6	H	A	$\underline{\text{A}}$	H	$\underline{\text{H}}$	A	H
7	H	A	H	$\underline{\text{H}}$	A	$\underline{\text{A}}$	H
8	H	A	H	A	$\underline{\text{A}}$	H	$\underline{\text{H}}$

immediately canceled out. Table 2.6 shows such a HAP for a league with $2n = 8$ teams. The HAP always consists of n pairs of complementary profiles for such schedules.

We may also consider the following problem that can be solved in polynomial time: given a timetable for the single round-robin tournament of a league with $2n$ teams, find an associated HAP where all teams have two consecutive breaks if one exists. The importance of this construction procedure is limited since it has been shown that such schedules can only exist if the league has 8 or 10 teams.

The requirement of consecutive breaks for each team can be modified as follows. Given a positive even integer k, a *k-interval* HAP is one in which each team has one home break and one away break, with the interval between them being exactly k (i.e., if the breaks occur in rounds s and s', then $|s' - s| = k$). For consecutive breaks, we have $k = 2$.

It has been shown that a k-interval HAP corresponding to a single round-robin timetable for $2n$ teams can be found only if $2k + 4 \leq 2n \leq 4k + 2$. If we assume that k is a constant, there is a polynomial algorithm running in time $O(n^{2k+4})$ that finds a k-interval HAP for a given timetable if one exists.

2.10 Travel Minimization and Break Maximization

Generally, the $2n$ teams of a league playing a single round-robin tournament are located in different cities, and the distances between the cities are known. All teams start the tournament at home, where they should return after their last away game. In this situation, minimizing the total distance traveled by the teams may be more critical than minimizing the number of breaks; see Sect. 1.3.2 for the distance minimization problem.

If all distances are assumed to be equal to one, then the problem becomes the minimization of the total number of travels the teams perform. A timetable T is given, and we want to find a HAP minimizing the total number of travels. We can show that the problem amounts to maximizing the number of breaks. Before the first round, the n teams starting with an away game will travel. At the end of the tournament, all n teams with an away game will travel back home. After each round $r = 1, \ldots, 2n-2$, only the teams with home breaks in the next round do not travel.

If $B(T)$ denotes the total number of breaks in the schedule based on timetable T, then the total number of home breaks is $B(T)/2$. Therefore, the total number of trips after round $2n-2$ is $2n(2n-2) - B(T)/2$. Hence, the total number of trips is

$$n + 2n(2n-2) - B(T)/2 + n = 2n(2n-1) - B(T)/2.$$

It follows that for minimizing the number of travels, we maximize the number $B(T)$ of breaks.

A schedule with the maximum number of breaks may be obtained from a schedule minimizing the number of breaks. To do so, we start from a HAP $\mathscr{H}$ with the minimum number of breaks and create another HAP $\mathscr{H}^*$ by exchanging the home-away status of all games in the even rounds of $\mathscr{H}$, as illustrated in Fig. 2.8.

Note that for every team i and every round $r = 2, \ldots, 2n-1$, entry (i, r) is a break in HAP $\mathscr{H}^*$ if and only if it is not a break in HAP $\mathscr{H}$. Hence, $B(\mathscr{H}) + B(\mathscr{H}^*) = 2n(2n-2)$ and $\mathscr{H}$ will be a HAP minimizing the number of breaks if and only if $\mathscr{H}^*$ is HAP maximizing the number of breaks.

2.11 Balanced Tournaments

We have assumed that each team has its venue (or stadium). This situation is not always the case, and we also have to consider models in which the $2n$ teams have to play their games in a set of shared venues. Special requirements regarding these venues (courts, fields, stadiums, locations) may also have to be considered for some sports scheduling problems.

We first consider a single round-robin tournament of a league with $2n$ teams and a collection of n venues. We would like to have a timetable as balanced as possible, i.e., in which all teams play approximately the same number of games in

Fig. 2.8 Construction of a HAP $\mathscr{H}^*$ with the maximum number of breaks from a HAP $\mathscr{H}$ with the minimum number of breaks for $2n = 6$ teams (breaks are underlined). (**a**) HAP $\mathscr{H}$ with $2n - 2 = 4$ breaks (minimum) indicated by underlined entries. (**b**) HAP $\mathscr{H}^*$ with $(2n-1)(2n-2) = 20$ breaks (maximum) indicated by underlined entries

	Rounds				
Teams	1	2	3	4	5
1	A	H	A	H	A
2	A	A	H	A	H
3	A	H	A	H	H
4	H	A	H	A	H
5	H	H	A	H	A
6	H	A	H	A	A

(a)

	Rounds				
Teams	1	2	3	4	5
1	A	A	A	A	A
2	A	H	H	H	H
3	A	A	A	A	H
4	H	H	H	H	H
5	H	A	A	A	A
6	H	H	H	H	A

(b)

each venue. Since each of the $2n$ teams has $2n-1$ games to play, one would like each team to have one or two games in each of the n venues. A schedule satisfying this requirement is called a *balanced tournament*. In some cases, instead of venues, the assigned time slots in the rounds (starting times of the games) have to be balanced, assuming that no games can be played in parallel. Balanced schedules exist for any $2n \neq 4$. A simple construction is described below for the cases where $2n = 0 \pmod 3$ or $2n = 2 \pmod 3$.

We start from a canonical ordered 1-factorization $(F_1, \ldots, F_{2n-1})$ of K_{2n}. For each round $r = 1, \ldots, 2n-1$, we denote by F_r^i the edge corresponding to the game scheduled at the venue $i = 1, \ldots, n$. In each round, games are assigned to venues as they are created by the circle method in Algorithm 1. For each round $r = 1, \ldots, 2n-1$, the game between teams $2n$ and r is first assigned to venue 1. Following, the game created in each iteration $\ell = 1, \ldots, n-1$ is assigned to venue $\ell + 1$.

At this point, all teams except team $2n$ have a balanced schedule because team $2n$ is always playing at the venue $i = 1$. For every $i = 1, \ldots, n-1$, we exchange the venues associated with edges F_i^1 and F_i^{i+1}. Similarly, for every $i = n+1, \ldots, 2n-1$, we exchange the venues associated with edges F_i^1 and F_i^{2n+1-i}. We observe that this construction balances the games of every team over the n venues, i.e., all teams play twice in all but one venue and once in the remaining venue. An illustration for a league of $2n = 6$ teams appears in Fig. 2.9.

Balanced tournaments for $2n$ teams have been shown to exist for all values of $2n \neq 4$. To make the problem more regular for a theoretical study, instead of

Fig. 2.9 Construction of a balanced tournament for $2n = 6$ teams (games in red had their venues changed). (**a**) Original unbalanced assignments of games to venues. (**b**) Balanced assignments of games to venues

	Venues		
Rounds	1	2	3
1	[6,1]	[2,5]	[3,4]
2	[6,2]	[3,1]	[4,5]
3	[6,3]	[4,2]	[5,1]
4	[6,4]	[5,3]	[1,2]
5	[6,5]	[1,4]	[2,3]

(a)

	Venues		
Rounds	1	2	3
1	[2,5]	[6,1]	[3,4]
2	[4,5]	[3,1]	[6,2]
3	[6,3]	[4,2]	[5,1]
4	[1,2]	[5,3]	[6,4]
5	[1,4]	[6,5]	[2,3]

(b)

using a complete graph K_{2n}, one may consider the graph K^*_{2n} obtained from K_{2n} by doubling all edges of a single 1-factor. We now have the following requirements:

(a) all pairs of teams meet at least once;
(b) there are n games in each round with $2n$ teams;
(c) exactly one game is played in each venue in each round;
(d) each team plays two games in each of the n venues; and
(e) each pair of teams plays no more than one game in each venue.

Finding a schedule meeting requirements (a) to (e) is slightly different from the balanced tournament design problem. For instance, although there is a balanced tournament for K_6, there are no schedules satisfying requirements (a) to (e) for K^*_6, as it may be seen by inspection. There is no solution either for K_4 or for K^*_4. However, a simple inductive construction was developed for K^*_{2n} when $2n$ is a power of two greater than or equal to eight. It is based on dividing the league with $2n$ teams into two disjoint sub-leagues. These ideas have been explored later to derive a procedure for the general case of $2n \geq 8$ teams.

In some situations, the number s of venues may be smaller than the number of games in each round of a compact schedule, i.e., $s < n$. A balanced non-compact schedule exists when $2n - 1 = 0 \bmod s$ or $n = 0 \bmod s$. As an illustration, consider a league with $2n = 10$ teams. Table 2.7 shows that a non-compact balanced schedule exists when the number of venues is $s = 3$. Note that $2n - 1 = 9$ is a multiple of $s = 3$. There are $n(2n - 1) = 45$ games to be scheduled in $n(2n - 1)/s = 15$ rounds.

In a more general setting, the number of available venues may vary from round to round. More precisely, we are given the number of rounds $q \geq 2n - 1$ and the number of venues s_r available in each round $r = 1, \ldots, q$. Balancing is no longer

Table 2.7 Balanced schedule for $2n = 10$ teams in $s = 3$ venues. Each team plays $2n - 1 = 9$ games, three of which in each venue

	Venues		
Rounds	1	2	3
1	[1,10]	[3,6]	[2,8]
2	[1,5]	[3,7]	[2,9]
3	[2,6]	[1,8]	[3,4]
4	[2,7]	[1,9]	[3,5]
5	[3,8]	[2,4]	[1,6]
6	[3,9]	[2,10]	[1,7]
7	[2,3]	[5,8]	[6,7]
8	[5,10]	[6,9]	[7,8]
9	[6,4]	[1,3]	[8,9]
10	[7,5]	[8,4]	[9,10]
11	[8,6]	[9,5]	[4,10]
12	[7,9]	[6,10]	[4,5]
13	[8,10]	[4,7]	[1,2]
14	[1,4]	[2,5]	[3,10]
15	[4,9]	[7,10]	[5,6]

an issue, but the question becomes to determine if a schedule exists where exactly s_r games are played in round $r = 1, \ldots, q$.

The condition $\sum_{r=1}^{q} s_r = n(2n - 1)$ holds and we assume that $s_r \leq n, r = 1, \ldots, q$. For convenience and without loss of generality, we assume that $s_1 \geq s_2 \geq \ldots \geq s_q$. This assumption does not restrict the model since the q rounds may be arbitrarily permuted.

A schedule with s_r games in each round $r = 1, \ldots, q$ always exists and can be constructed from a canonical 1-factorization of K_{2n}. For each 1-factor F_i of the ordered canonical 1-factorization $\mathscr{F} = (F_1, \ldots, F_{2n-1})$, we pick each of its edges in the order $F_i^1, \ldots, F_i^n$ and assign it to the first round with a place still available. Figure 2.10 illustrates this construction for $2n = 8$ teams on ten rounds. The number of venues available in each round is $s_1 = s_2 = s_3 = 4$, $s_4 = s_5 = s_6 = 3$, $s_7 = s_8 = s_9 = 2$, and $s_{10} = 1$. This strategy gives a schedule where each team plays at most one game in each round. Figure 2.10a gives the ordered canonical 1-factorization of K_8. Figure 2.10b displays the games scheduled in each round.

2.12 Tournaments with Referees

We now consider another type of scheduling problem related to the assignment of referees to games. We want to construct a timetable for a single round-robin tournament of a league with $2n$ teams for which we have $2n - 1$ referees. Exactly one referee should be assigned to each game. The goal is to find a compact timetable in $2n - 1$ rounds such that each team sees every referee exactly once.

Fig. 2.10 Construction of a timetable with s_r venues in round r. (**a**) Ordered canonical 1-factorization of K_8. (**b**) Games scheduled in each round

	Edges			
1-factors	1	2	3	4
F_1	[8,1]	[2,7]	[3,6]	[4,5]
F_2	[8,2]	[3,1]	[4,7]	[5,6]
F_3	[8,3]	[4,2]	[5,1]	[6,7]
F_4	[8,4]	[5,3]	[6,2]	[7,1]
F_5	[8,5]	[6,4]	[7,3]	[1,2]
F_6	[8,6]	[7,5]	[1,4]	[2,3]
F_7	[8,7]	[1,6]	[2,5]	[3,4]

(a)

	Games			
Rounds	1	2	3	4
1	[8,1]	[2,7]	[3,6]	[4,5]
2	[8,2]	[3,1]	[4,7]	[5,6]
3	[8,3]	[4,2]	[5,1]	[6,7]
4	[8,4]	[5,3]	[6,2]	
5	[7,1]	[8,5]	[6,4]	
6	[7,3]	[1,2]	[8,6]	
7	[7,5]	[1,4]		
8	[2,3]	[8,7]		
9	[1,6]	[2,5]		
10	[3,4]			

(b)

The combinatorial structure needed to model this problem is a *Room square* (of size $2n-1$) built on a set $T = \{1, \ldots, 2n\}$ corresponding to the $2n$ teams. A Room square is a $(2n-1) \times (2n-1)$ array where every entry is either empty or contains an unordered pair of distinct elements $i, j \in T$, corresponding to the game between teams i and j. In addition, it satisfies the following conditions: (1) every element of T occurs exactly once in every row and exactly once in every column, and (2) every unordered pair of elements of T appears in exactly one entry of the Room square.

An example of a Room square of size $2n-1 = 7$ is shown in Fig. 2.11. There are eight teams, and the timetable is represented by the nonempty entries of the Room square. Its rows correspond to the $2n-1$ rounds, and its columns are associated with the $2n-1$ referees. Room squares can be found for any odd number greater than or equal to seven.

We may now consider a further extension of the problem with $2n-1$ referees by introducing $2n-1$ venues. We require, in addition, that each team plays its $2n-1$ games in different venues.

This extension introduces a third dimension to the Room square previously considered. We now have a *Room cube* of size $2n-1$, i.e., a 3-dimensional array with the property that each one of its 2-dimensional projections is a Room square of size $2n-1$.

A Room cube based on the Room square of Fig. 2.11 is shown in Fig. 2.12. Each non-empty entry in row r and column u of the Room cube contains a pair $[i, j]$ and

Referees

	[8, 1]	[4, 5]	[2, 7]		[3, 6]		
		[8, 2]	[5, 6]	[1, 3]		[4, 7]	
			[8, 3]	[6, 7]	[2, 4]		[1, 5]
Rounds	[2, 6]			[8, 4]	[7, 1]	[3, 5]	
		[3, 7]			[8, 5]	[1, 2]	[4, 6]
	[5, 7]		[4, 1]			[8, 6]	[2, 3]
	[3, 4]	[6, 1]		[2, 5]			[8, 7]

Fig. 2.11 Room square of size $2n - 1 = 7$

Referees

	1 [8 ,1]	7 [4, 5]	6 [2, 7]		4 [3, 6]		
		2 [8, 2]	1 [5, 6]	7 [1, 3]		5 [4, 7]	
			3 [8, 3]	2 [6, 7]	1 [2, 4]		6 [1, 5]
Rounds	7 [2, 6]			4 [8, 4]	3 [7, 1]	2 [3, 5]	
		1 [3, 7]			5 [8, 5]	4 [1, 2]	3 [4, 6]
	4 [5, 7]		2 [4, 1]			6 [8, 6]	5 [2, 3]
	6 [3, 4]	5 [6, 1]		3 [2, 5]			7 [8, 7]

Fig. 2.12 Room cube of size $2n - 1 = 7$

a venue identification s, meaning that teams i and j play against each other in round r with referee u at the venue s.

2.13 Carry-Over Effects

We have introduced the carry-over effects in Sect. 1.3.5. Minimizing the value of the carry-over effects is another requirement often present in the construction of schedules for sports tournaments.

In this problem, we seek to avoid having many teams such that all of them play against two teams, i and j, consecutively in the same order. In other words, if we consider the sets formed by ordered pairs (i, j) of consecutive opponents of each team, we would like to have all of them pairwise disjoint. Otherwise, if this is

Fig. 2.13 Timetable with no repeating pairs of consecutive opponents. (**a**) Opponents of teams 1, 2, 3, and 4 in each round of a cyclic timetable. (**b**) Pairs of consecutive opponents of teams 1, 2, 3, and 4. (**c**) Two orthogonal Latin squares obtained from the above pairs of consecutive opponents

	Rounds			
Teams	1	2	3	1
1	4	3	2	4
2	3	4	1	3
3	2	1	4	2
4	1	2	3	1

(a)

Team i (i, i)	Pairs of consecutive opponents of team i		
(1,1)	(4,3)	(3,2)	(2,4)
(2,2)	(3,4)	(4,1)	(1,3)
(3,3)	(2,1)	(1,4)	(4,2)
(4,4)	(1,2)	(2,3)	(3,1)

(b)

1	4	3	2
2	3	4	1
3	2	1	4
4	1	2	3

1	3	2	4
2	4	1	3
3	1	4	2
4	2	3	1

(c)

impossible, we want to minimize the sum of the squares of the number of times each pair of consecutive opponents appears over all sets.

We define a *Latin square* as a $q \times q$ array in which every row contains the numbers 1 to q, and so does every column.

The timetable in Fig. 2.13 illustrates an example of a league with $2n = 4$ teams. Figure 2.13a gives the opponents of teams 1 to 4 in each round. For convenience, we consider a cyclic timetable, assuming that round 1 follows (or is repeated after) round 3. Figure 2.13b displays the pairs of consecutive opponents for each team. Team 1, for example, meets consecutively with teams 4, 3, and 2. Hence, the pairs of consecutive opponents for team 1 in the cyclic schedule are (4,3), (3,2), and (2,4). We artificially introduce the pair (i, i) in the first column of the table of ordered pairs of consecutive opponents (although it does not correspond to a real pair of opponents). This table can be seen as a $2n \times 2n$ array containing $4n^2$ pairs of teams. We would like to have (if possible) all these pairs different, as it happens to be the case in Fig. 2.13b. If we decompose the $2n \times 2n$ array of ordered pairs into two $2n \times 2n$ arrays by taking the first element i (resp. the second element j) of each ordered pair (i, j), we obtain two Latin squares. Requiring that all ordered pairs differ amounts to saying that the two Latin squares are *orthogonal*. Since orthogonal Latin squares of size six do not exist, in any timetable for a league with $2n = 6$ teams, at least two teams will consecutively meet some pair of teams. Efforts have been devoted to constructing schedules where pairs of consecutive opponents are as different as possible; see Sect. 5.1.

It has nevertheless been shown that if $2n$ is a power of two, then there is a schedule in $2n - 1$ rounds such that each pair of consecutive opponents occurs precisely once.

It should be noted that the canonical 1-factorization of K_{2n} maximizes the value of the carry-over effects. This observation reinforces the need to examine other 1-factorizations, not isomorphic to the canonical 1-factorization; see Sect. 2.16.

2.14 List-Edge Coloring

We have seen that constructing a timetable for a sports tournament can be formulated as an edge coloring problem in a complete graph K_{2n}.

Let us consider an extension of this model by taking a graph $G = (V, E)$ and a set $C = \{1, \ldots, q\}$ of colors. For each edge $e \in E$, we have a list $L(e) \subseteq C$ of feasible colors. A list coloring $\mathscr{L}$ of (the edges of) G is an edge coloring $\{F_1, \ldots, F_q\}$, where each $F_r, r = 1, \ldots, q$, is a matching and each edge $e \in E$ gets some color $c(e) \in L(e)$. This model applies to situations where some games are constrained to be played in specific rounds.

This model is used frequently in school timetabling, where each lecture $e = [c_i, t_j]$ given by a teacher t_j to a class c_i may take place only at some specific periods belonging to $L(e)$. Finding a timetable amounts to determining a list-edge coloring of the (bipartite) graph $G = (D \cup T, E)$ in which D (resp. T) is the set of nodes associated with the classes (resp. teachers) and E is the collection of lectures to be scheduled.

When scheduling a single round-robin tournament of a league with $2n$ teams, we have a list-edge coloring problem in the case where a HAP is given, and one has to find a timetable associated with this HAP. This strategy is the first-break, then-schedule decomposition. Determining the ordered 1-factorization $(F_1, \ldots, F_{2n-1})$ is a restricted edge coloring problem. To consider the given HAP, for any pair of teams i and j, the game between them can only be played in rounds where one plays a home game and the other an away game. In terms of list-edge coloring, for each edge $e = [i, j]$ of K_{2n}, the list $L(e)$ is formed by the rounds in which i is at home, and j is away or vice-versa.

In fact, the integer programming formulation (2.2)–(2.6) presented in Sect. 2.7 is a formulation for the list-edge coloring problem maximizing the number of edges that may be colored. The variable $x_{ijr} = 1$ if the edge $[i, j]$ gets the color r. In this case, $L(e = [i, j]) = \{r = 1, \ldots, 2n - 1 : \ h_{ir} \neq h_{jr}\}$.

2.15 Exploring the Landscape of Edge Colorings

Edge coloring and list-edge coloring are often used for scheduling problems in round-robin tournaments. We have seen that canonical 1-factorizations of K_{2n}

are the most popular construction of timetables. However, the existence of many additional requirements will compel us to explore the massive set of 1-factorizations to obtain non-canonical 1-factorizations satisfying the additional constraints that may be imposed.

A widely used technique for modifying a 1-factorization $\{F_1, \ldots, F_{2n-1}\}$ is that of *bichromatic exchanges*. Given two 1-factors F_a and F_b, consider the set formed by all edges of $F_a \cup F_b$. It consists of a collection of node disjoint chains and cycles whose edges are colored alternately with colors a and b. These are called *alternating cycles* and *alternating chains*.

Each $F_a \cup F_b$ consists exclusively of even alternating cycles. Exchanging colors a and b in $F_a \cup F_b$ will not give a different 1-factorization: in terms of tournament scheduling, we simply exchange the games of round a with those of round b.

However, if $F_a \cup F_b$ has more than one connected component, then exchanging colors exclusively inside one of these components and leaving the others unchanged will give a different 1-factorization. This idea can be explored to generate different timetables in an edge recoloring procedure that searches for a timetable satisfying specific constraints.

Bichromatic exchanges are sometimes unable to reach some 1-factorizations from an arbitrary initial 1-factorization. Consider for example the 3-regular bipartite graph illustrated in Fig. 2.14 and the two different 1-factorizations $\{F_1, F_2, F_3\}$ in Fig. 2.14a and $\{F'_1, F'_2, F'_3\}$ in Fig. 2.14b. We note that $F_a \cup F_b$ and $F'_a \cup F'_b$ are connected for every $a \neq b$, with $a, b = 1, 2, 3$. Hence, bichromatic exchanges will not transform the first 1-factorization into the second, or conversely.

Therefore, it is necessary to consider more elaborate edge recoloring procedures that involve more than two colors. One of these procedures can be described as follows. Take two nodes v_1 and v_2 of the graph K_{2n} with a 1-factorization $\{F_1, \ldots, F_{2n-1}\}$. Consider a collection of edge-disjoint alternating chains $C_1, \ldots, C_p$ of even length between v_1 and v_2, such that the set $\mathcal{S}$ of colors occurring on the edges of the chains adjacent to v_1 is the same as the set of colors occurring on the edges of the chains adjacent to v_2. Notice that such a collection of

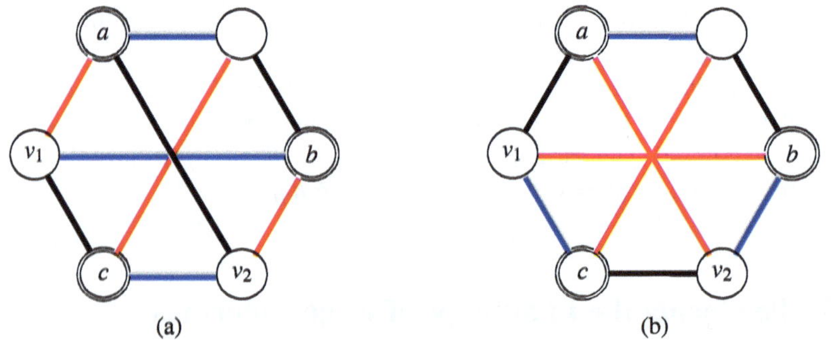

Fig. 2.14 Two 1-factorizations of a complete bipartite 3-regular graph (1-factors in blue, red, and black). (**a**) First 1-factorization: $\{F_1, F_2, F_3\}$. (**b**) Second 1-factorization: $\{F'_1, F'_2, F'_3\}$

chains always exists in K_{2n}: we may take, e.g., $2n-2$ chains of length two between v_1 and v_2. By interchanging the two colors in each of the chains $C_1, \ldots, C_p$, we obtain a new edge coloring that is different from the initial edge coloring.

To illustrate this edge recoloring procedure, once again, we consider the 3-regular complete bipartite graph of Fig. 2.14 with the 1-factorization $\{F_1, F_2, F_3\}$ in Fig. 2.14a. We take three alternating chains of length two between v_1 and v_2: $C_1 =< v_1, a, v_2 >$, $C_2 =< v_1, b, v_2 >$, and $C_3 =< v_1, c, v_2 >$. The set of colors occurring in the edges of the chains incident to both v_1 and v_2 is $S =$ {blue, red, black}. Interchanging the colors red and black in C_1, colors blue and red in C_2, and colors black and blue in C_3, we obtain the 1-factorization $\{F_1', F_2', F_3'\}$ in Fig. 2.14b, which could not be reached from $\{F_1, F_2, F_3\}$ by using only bichromatic exchanges.

More involved edge recoloring procedures may generate different 1-factorizations from the same initial 1-factorization. This kind of procedure will be frequently used for better exploring the set of 1-factorizations of K_{2n} by repeatedly moving from one 1-factorization to another.

The systematic use of these and other edge recoloring procedures in the search for better schedules is the subject of Chap. 3.

2.16 How to Find General 1-Factorizations?

The chromatic index $\chi'(G)$ of a graph $G = (V, E)$ was defined in Sect. 2.1. Vizing's theorem states that if G is a simple graph (i.e., a graph without loops or parallel edges), then its chromatic index $\chi'(G) \leq \Delta(G) + 1$, where $\Delta(G)$ is the largest degree of any node of graph G. Different constructive proofs of Vizing's theorem provide algorithms for constructing edge colorings of G using, at most, $\Delta(G) + 1$ colors.

In the particular case of complete graphs, $\Delta(K_{2n}) = 2n - 1$ and $\Delta(K_{2n-1}) = 2n-2$. Vizing's theorem implies that $\chi'(K_{2n}) \leq \Delta(K_{2n})+1 = 2n$ and $\chi'(K_{2n-1}) \leq \Delta(K_{2n-1}) + 1 = 2n - 1$. It turns out that this bound is tight for K_{2n-1}, but not for K_{2n} because, as we have already seen, K_{2n} can be colored with $2n - 1$ colors.

An edge coloring of K_{2n} with $2n - 1$ colors (i.e., a 1-factorization of K_{2n}) can be obtained from an edge coloring of K_{2n-1} with the same $2n - 1$ colors that, in turn, can be built using the constructions of Vizing's theorem. If K_{2n-1} can be edge colored with $2n - 1$ colors, it is possible to obtain an edge coloring with the same number of colors for K_{2n}. First, we observe that a different color (among the $2n - 1$ colors) is missing at the edges incident to each node of K_{2n-1}. Since each color is used in exactly $n - 1$ edges of K_{2n-1}, it is missing in one node. Therefore, we can add a new node to the graph K_{2n-1} and color the new edges connecting this node to the original nodes of K_{2n-1} with the missing color of each node. Hence, an edge coloring of K_{2n} with $2n - 1$ colors is obtained. An application of this procedure is illustrated in Fig. 2.15.

We now show that the edges of an arbitrary graph G can be colored with $\Delta(G)+1$ colors. The procedure starts from a graph G partially colored with no more than

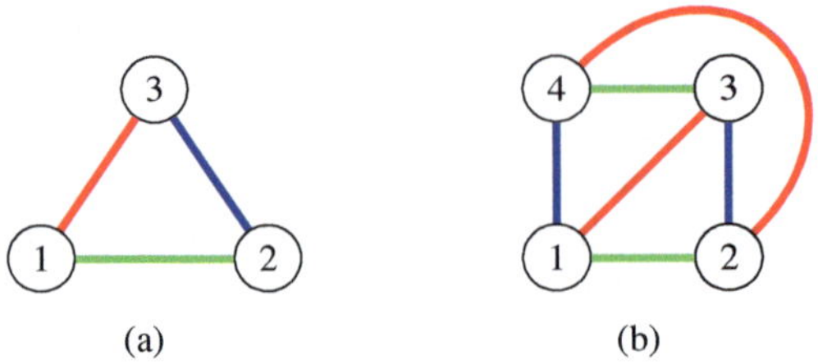

Fig. 2.15 Finding an edge coloring of K_4 (right) from an edge coloring of K_3 (left). (**a**) Edge coloring of K_3: $\chi'(K_3) = 3$. (**b**) Edge coloring of K_4: $\chi'(K_4) = 3$

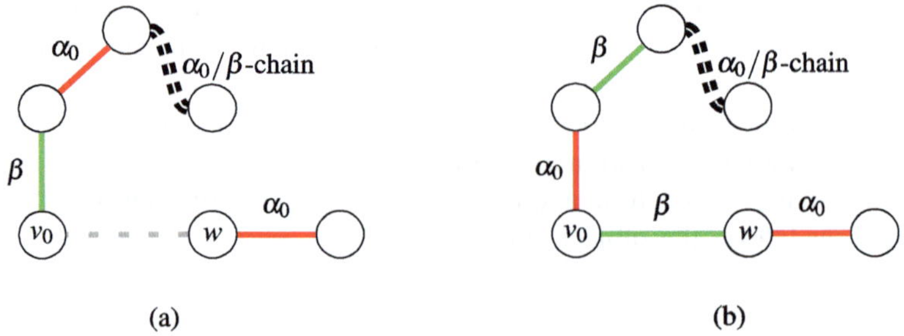

Fig. 2.16 First case of $Free(w) \cap Free(v_0) = \emptyset$: color β becomes free at vertex v_0 after the color exchange along the α_0/β-chain. (**a**) Before. (**b**) After

$\Delta(G) + 1$ colors. At each step, a new, yet uncolored edge $e_0 = [w, v_0]$ will be colored without using more than $\Delta(G) + 1$ colors. This procedure is repeated until all edges are colored when it obtains the desired edge coloring with $\Delta(G)+1$ colors.

We denote by $Free(v)$ the set of colors missing at node v, i.e., the set of colors not used in the edges incident to node v. Since the degrees of nodes v_0 and w are smaller than or equal to $\Delta(G)$, we have $|Free(v_0)| \geq 2$ and $|Free(w)| \geq 2$.

If there is a color c that belongs to $Free(v_0) \cap Free(w)$, then we simply color edge e_o with c.

Otherwise, $Free(w) \cap Free(v_0) = \emptyset$. Then, let $\alpha_0 \in Free(v_0)$ and $\beta \in Free(w)$ be two free colors at nodes v_0 and w, respectively. Let P be a maximal chain starting from v_0 such that its edges are alternately colored with colors β and α_0. Such a chain will be called an α_0/β-chain. Two cases may arise: P does not end in node w or P ends in node w. In the first case, edge $e_0 = [w, v_0]$ may be colored, and the edge coloring is augmented. This is done by exchanging the colors of the edges along the chain P and coloring edge $e_0 = [w, v_0]$ with color β that is now free in v_0. This situation is illustrated in Fig. 2.16.

In the second case, we assume that the chain P ends at node w. If we exchange the colors of all edges of P, then $\beta \in Free(v_0)$ but $\beta \notin Free(w)$, as illustrated in Fig. 2.17.

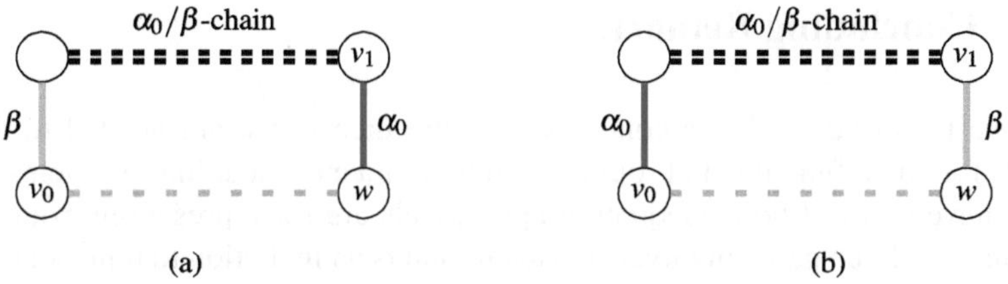

Fig. 2.17 Second case of $Free(w) \cap Free(v_0) = \emptyset$: path P ends at vertex w. On the left, $\beta \notin Free(v_0)$ and $\beta \in Free(w)$ before the color exchanges. On the right, $\beta \in Free(v_0)$ and $\beta \notin Free(w)$ after the color exchanges. (**a**) Before. (**b**) After

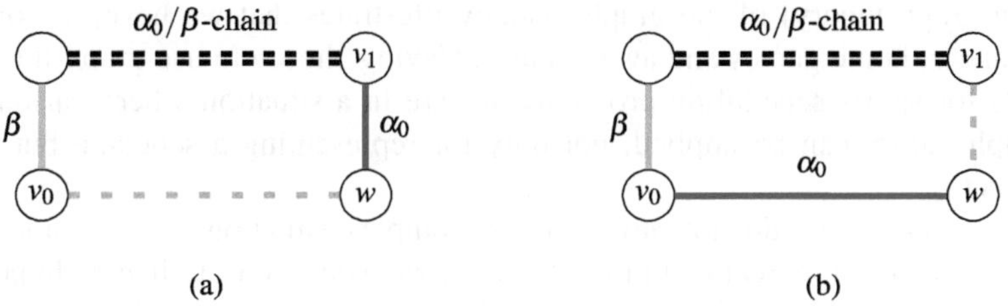

Fig. 2.18 Second case of $Free(w) \cap Free(v_0) = \emptyset$: path P ends at vertex w. On the right, color α_0 moves from edge $[w, v_1]$ to edge $[w, v_0]$. (**a**) Before. (**b**) After

Then, let v_1 be the node adjacent to w in P and $e_1 = [w, v_1]$ be the edge that is colored with α_0. At this point, the color α_0 is removed from edge e_1 and edge e_0 is colored with that same color α_0, as illustrated in Fig. 2.18.

The problem consists of recoloring edge e_1. The same procedure applied to edge e_0 can be used to assign a color to edge e_1. If $Free(w) \cap Free(v_1)$ is also empty, then the color to be selected from $Free(v_1)$ must be different from α_0 in order to avoid cycling. This color always exists, since $|Free(v_1)| \geq 2$.

We may continue with this procedure until $Free(w) \cap Free(v_i)$ becomes not empty or the chain P does not terminate at w. This case will eventually occur after, at most, $\Delta(G)$ iterations. This proof of Vizing's theorem immediately yields a polynomial algorithm running in time $O(|E||V|\Delta(G))$ to obtain a proper edge $(\Delta(G) + 1)$-coloring of a simple graph. More efficient algorithms have been designed and used in practical instances.

This edge-coloring procedure can be applied to a complete graph, in particular to K_{2n-1} to obtain an edge coloring with $2n - 1$ colors, from which an edge coloring of K_{2n} with the same number of colors can be obtained. It will provide a 1-factorization of graph K_{2n}.

Remark 2.5 Given any arbitrary 1-factorization of K_{2n}, there exists an order for coloring its edges by the procedure described above that will produce exactly this 1-factorization. Contrarily to the circle method, this procedure may generate any 1-factorization of K_{2n}.

2.17 Concluding Remarks

We have reviewed the basic concepts of graph theory that are needed to develop models for some fundamental round-robin tournament scheduling problems.

We have realized how adequate graph models are for representing round-robin tournaments. Each game involves two teams and is an invitation to represent it by an edge linking two nodes. Complete graphs also appear very clearly whenever every two teams have to play a game against each other. Since a team can not play more than one game in a round, we come across 1-factorizations to represent timetables.

In addition, we observe that assigning orientations to the edges (a fundamental idea in graph theory) of the graph visually illustrates that we have, in some circumstances, home games and away games. Having chosen to use graph-theoretical models for sports scheduling problems, we are in a situation where various tools of graph theory can be applied, not only for representing a schedule but also to construct it.

Conveniently, we do not have a more complex situation where games could involve three or more teams. In this situation, we would have to handle hyperedges (defined, e.g., by triples or quadruples of nodes), and we would have to handle hypergraphs instead of graphs. The trouble is that besides some exceptional cases, we only have a few efficient algorithms to tackle hypergraph coloring problems. Thus we could, fortunately, stick to graph models.

We will show later that the role of graphs will also provide a deeper understanding of the techniques used for exploring the set of 1-factorizations or, more generally, that of edge colorings.

2.18 Related Literature

Graphs are the essential ingredients of all models described in this chapter. Section 2.1 recalls some basic concepts and definitions of graph theory. For more definitions, results, and applications of graph theory and algorithms, we refer the reader to Berge (1985), Gould (1988), or Diestel (2010). An excellent presentation of discrete mathematics structures related to tournament construction can be found in Anderson (2006).

Section 2.2 introduces the basic graph model used to represent single round-robin tournaments. Properties and variations of edge colorings are found in Fiorini and Wilson (1977), Folkman and Fulkerson (1969), and Gabow et al. (1985). One-factorizations, a particular case of edge colorings, are discussed in Akiyama and Kano (1985) and Mendelsohn and Rodney (1994). The circle method was proposed originally by Kirkman (1847). Goossens and Spieksma (2012) examined the popularity of the canonical schedule and how it was used in 25 European football (soccer) competitions for the season 2008-2009. The authors also discussed how the schedules managed the carry-over effects. Lambrechts et al. (2018) showed that schedules generated for round-robin tournaments by the circle method have maximum carry-over effects.

Section 2.3 discusses home and away games, showing how we move from timetables to schedules via the creation of home-away patterns (HAPs). Using graphs to model fundamental sports scheduling problems with HAPs was first presented in de Werra (1980, 1981, 1982).

The 1-factorization described in Sect. 2.4 was motivated by an application to an ice hockey league in Switzerland described in de Werra (1980). The players were semi-professional, and the schedule should avoid long trips during weekdays. The schedule structure was such that all games were played alternately on weekdays and weekends. This construction can be extended to the case where the league with $2n$ teams is divided into an even number of subleagues of the same size. The procedures are given in de Werra (1985).

Section 2.5 deals with multi-round-robin tournaments and, in particular, with double round-robin tournaments. Specific constructions must be set up to avoid additional breaks when concatenating several tournaments. These constructions appeared in de Werra (1981).

The structure of HAPs with a minimum number of breaks is discussed in Sect. 2.6. Procedures for finding a HAP with minimum breaks starting from a given timetable appeared in Miyashiro and Matsui (2005). Post and Woeginger (2006) studied the worst cases that can happen when minimizing the number of breaks of a given schedule. They gave upper bounds on the minimum number of breaks. They showed that break minimization for a given partial timetable with $2n$ teams and only three rounds is NP-hard.

Section 2.7 deals with the construction of schedules from HAPs or timetables. An integer programming model was devised in Briskorn (2008) for constructing timetables from a given HAP. It is a list-edge-coloring model in which one maximizes the number of edges that can be colored. Necessary (but not sufficient) conditions for the existence of a timetable compatible with a given HAP were also proposed. A stronger conjecture was disproved by Horbach (2010). Miyashiro et al. (2003) also proposed necessary conditions.

Section 2.8 considers the minimization of the number of rounds where breaks occur. It amounts to partitioning the edge set of K_{2n} into a minimum number of bipartite graphs. Related results and properties, including the proof of Proposition 2.8, are found in Kotzig (1972) and de Werra (1982).

Fairness of schedules, discussed in Sect. 2.9, has generated much work. Knust and von Thaden (2006) considered the situation where HAPs have to be constructed in such a way that, for each team, the numbers of home and away games differ by at most one. They suggested repairing techniques for improving unbalanced HAPs. Tanaka and Miyashiro (2016) studied HAPs for which each team has one home break and one away break. They require that the distance between the two breaks is the same for all teams, and properties of such fair schedules were derived. Suzuka et al. (2007) also examined fair schedules in terms of breaks.

Some connections between break minimization and traveling tournaments discussed in Sect. 2.10 were presented in Urrutia and Ribeiro (2006).

Balanced tournaments is the subject of Sect. 2.11. They were defined and studied in Schellenberg et al. (1977) and occur when n venues are available for a single round-robin tournament of a league with $2n$ teams. A regularized version of the problem was studied by de Werra et al. (2006) and Ikebe and Tamura (2008). In Mendelsohn and Rodney (1994), stadium-balanced tournament design was studied for an arbitrary number of stadiums. Geinoz et al. (2008) studied some constructions of balanced schedules based on partitions into subleagues. Erzurumluoglu (2018) exploited the idea of partitioning the league for constructing several types of balanced schedules.

Tournaments with referees are considered in Sect. 2.12. The inclusion of referees in the models, together with some balancing requirements, led to the use of Room squares, whose original concept appeared in Kirkman (1850) and was later formalized by Room (1955) and studied by Mullin and Wallis (1975). Construction procedures for Room squares appeared in Anderson (1997). Their extensions to Room cubes and p-dimensional Room cubes are presented in Dinitz and Stinson (2005).

The carry-over effect discussed in Sect. 2.13 has generated extensive research due to the irregularities it may create in a tournament. It was studied first in Russell (1980), where a construction for a balanced tournament was devised for the case when the number of teams is a power of two. Lambrechts et al. (2018) showed that the circle method (which gives a canonical 1-factorization) maximizes the carry-over effects for single round-robin tournaments, which should be a reason to use other types of 1-factorizations. Heuristics for the minimum carry-over effects problem appeared, e.g., in Guedes and Ribeiro (2011) and Cao et al. (2022).

List-edge coloring is the subject of Sect. 2.14. Some related questions have been discussed in Galvin (1995) and Borodin et al. (1997).

Section 2.15 describes techniques for modifying edge colorings, which are adequate tools to explore other 1-factorizations. They have been presented, in particular, in de Werra (1971, 1988), Januario and Urrutia (2016), and Januario et al. (2016).

Section 2.16 presents the method of Vizing (1964) to find an initial 1-factorization. This method, based on edge-coloring procedures, gives the well-known upper bound on the chromatic index of a graph. Gould (1988) also gave a simple proof of this result. Gabow et al. (1985) presented another edge-coloring procedure that can be used for constructing 1-factorizations.

References

Akiyama J, Kano M (1985) Factors and factorizations of graphs - a survey. J Graph Theory 9:1694–1710

Anderson I (1997) Combinatorial designs and tournaments. Oxford lecture series in mathematics and its applications. Oxford University Press, Oxford

Anderson I (2006) Combinatorial designs and tournaments. Oxford lecture series in mathematics and its applications, 2nd edn. Oxford University Press, Oxford

Berge C (1985) Graphs and hypergraphs. Elsevier, Amsterdam
Borodin OV, Kostochka AV, Woodall DR (1997) List edge and list total colourings of multigraphs. J Comb Theory Ser B 71:184–204
Briskorn D (2008) Feasibility of home-away-pattern sets for round robin tournaments. Oper Res Lett 36:283–284
Cao Y, Wu W, Yagiura M (2022) A metaheuristic algorithm for the carry-over effect value minimization problem in round-robin tournaments. J Adv Mech Design Syst Manuf 16(4). https://doi.org/10.1299/jamdsm.2022jamdsm0042
Diestel R (2010) Graph theory, graduate texts in mathematics, vol 173, 4th edn. Springer, Berlin
Dinitz JH, Stinson DR (2005) On assigning referees to tournament schedules. Bull Inst Comb Appl 44:22–28
Erzurumluoglu A (2018) Constructive day-balanced round-robin tournaments with partitions. Discrete Appl Math 235:81–91
Fiorini S, Wilson RJ (1977) Edge-colourings of graphs. Pitman, Wetherby
Folkman J, Fulkerson DR (1969) Edge-colorings in bipartite graphs. In: Combinatorial mathematics and its applications. University of North Carolina, Chapel Hill, Press, pp 561–578
Gabow HN, Nishizeki T, Kariv DOL, Terada O (1985) Algorithms for edge-coloring graphs. Technical Report TRECIS-8501, Tohoku University
Galvin F (1995) The list chromatic index of a bipartite multigraph. J Comb Theory Ser B 63:153–158
Geinoz A, Ekim T, de Werra D (2008) Construction of balanced sports schedules using partitions into subleagues. Oper Res Lett 36:279–282
Goossens DR, Spieksma FCR (2012) Soccer schedules in Europe: an overview. J Sched 15:641–651
Gould RJ (1988) Graph theory. Benjamin Cummings, San Francisco
Guedes ACB, Ribeiro CC (2011) A heuristic for minimizing weighted carry-over effects in round robin tournaments. J Sched 14:655–667
Horbach A (2010) A combinatorial property of the maximum round robin problem. Oper Res Lett 38:121–122
Ikebe Y, Tamura A (2008) On the existence of sports schedules with multiple venues. Discrete Appl Math 156:1694–1710
Januario T, Urrutia S (2016) A new neighborhood structure for round robin scheduling problems. Comput Oper Res 70:127–139
Januario T, Urrutia S, Ribeiro CC, de Werra D (2016) Edge coloring: a natural model for sports scheduling. Eur J Oper Res 254:1–8
Kirkman TP (1847) On a problem in combinations. Camb Dublin Math J 2:191–204
Kirkman TP (1850) Note on an unanswered prize question. Camb Dublin Math J 5:255–262
Knust S, von Thaden M (2006) Balanced home-away assignments. Discrete Optim 3:354–365
Kotzig A (1972) Decompositions of complete graphs into regular bichromatic factors. Discrete Math 2:383–387
Lambrechts E, Ficker AMC, Goossens DR, Spieksma FCR (2018) Round-robin tournaments generated by the circle method have maximum carry-over. Math Program 172:277–302
Mendelsohn E, Rodney P (1994) The existence of court balanced tournament designs. Discrete Math 133:207–216
Miyashiro R, Matsui T (2005) A polynomial-time algorithm to find an equitable home-away assignment. Oper Res Lett 33:235–241
Miyashiro R, Iwasaki H, Matsui T (2003) Characterizing feasible pattern sets with a minimum number of breaks. In: Burke EK, de Causmaecker P (eds) Practice and theory of automated timetabling IV. Lecture notes in computer science, vol 2740. Springer, Berlin, pp 78–99
Mullin RC, Wallis WD (1975) The existence of Room squares. Aequ Math 13:1–7
Post G, Woeginger GJ (2006) Sports tournaments, home-away assignments, and the break minimization problem. Discrete Optim 3:165–173
Room TG (1955) A new type of magic square. Math Gazette 39:307–307

Russell KG (1980) Balancing carry-over effects in round robin tournaments. Biometrika 67:127–131

Schellenberg P, van Rees G, Vanstone S (1977) The existence of balanced tournament designs. Ars Comb 3:303–318

Suzuka A, Miyashiro R, Yoshise A, Matsui T (2007) The home-away assignment problems and break minimization/maximization problems in sports scheduling. Pac J Optim 3:113–133

Tanaka T, Miyashiro R (2016) Fair home-away tables in sports scheduling. J Adv Mech Design Syst Manuf 10(3):JAMDSM0037

Urrutia S, Ribeiro CC (2006) Maximizing breaks and bounding solutions to the mirrored traveling tournament problem. Discrete Appl Math 154:1932–1938

Vizing VG (1964) On an estimated of the chromatic class of a p-graph. Diskrete Analiz 3:25–30. In Russian

de Werra D (1971) Equitable colorations of graphs. Revue Française d'Informatique et de Recherche Opérationnelle 5:3–8

de Werra D (1980) Geography, games, and graphs. Discrete Appl Math 2:327–337

de Werra D (1981) Scheduling in sports. In: Hansen P (ed) Studies on graphs and discrete programming. North Holland, Amsterdam, pp 381–395

de Werra D (1982) Minimizing irregularities in sports schedules using graph theory. Discrete Appl. Math 4:217–226

de Werra D (1985) On the multiplication of divisions: the use of graphs for sports scheduling. Networks 15:125–136

de Werra D (1988) Some models of graphs for scheduling sports competitions. Discrete Appl Math 21:47–65

de Werra D, Ekim T, Raess C (2006) Construction of sports schedules with multiple venues. Discrete Appl Math 154:47–58

Chapter 3
Metaheuristics and Local Search

Combinatorial optimization, which includes sports scheduling problems as one of its areas of application, consists in finding optimal solutions to problems defined over a discrete set of feasible solutions.

Any combinatorial optimization problem can be formulated as the constrained minimization (resp. maximization) of some function $f(S) : S \in F$. Here, E is a discrete ground set formed by the elements that form the problem solutions, $F \subseteq 2^E$ is the set of feasible solutions, and $f : 2^E \rightarrow \mathbb{R}$ is the objective function. Any solution S for this problem is formed by a subset of the elements of the ground set. We seek an *optimal solution* (or, simply, an *optimum*) $S^* \in F$ such that $f(S^*) \leq f(S)$ (resp. $f(S^*) \geq f(S)$), for all $S \in F$. The ground set E, the objective function f, and the feasible set F are specific to each problem instance.

Even with the advent of new computer technologies and parallel processing, many of these problems can not be solved to optimality in reasonable computation times due to their intrinsic complexity or to their size. Furthermore, reaching optimal solutions is meaningless in many practical situations since we are often dealing with rough simplifications of reality and the available data is not precise. The goal of approximate algorithms (or heuristics) is to quickly produce good suboptimal solutions that do not guarantee optimality.

Metaheuristics are general high-level procedures that coordinate simple methods and rules to find high-quality (often optimal) approximate solutions to hard optimization problems. Among them, we find simulated annealing, tabu search, greedy randomized adaptive search procedures (GRASP), variable neighborhood search (VNS), iterated local search (ILS), genetic algorithms, scatter search, and others. They have been applied to various academic and real-world problems, particularly sports scheduling.

This chapter overviews approximate optimization methods, from greedy and semi-greedy construction algorithms to local search and from local search to metaheuristics. Some fundamental principles of metaheuristics are reviewed, and their templates are provided. Several approaches to sports scheduling problems make use

C. C. Ribeiro et al., *Combinatorial Models for Scheduling Sports Tournaments*,
EURO Advanced Tutorials on Operational Research,
https://doi.org/10.1007/978-3-031-37283-4_3

of local search in the development of heuristics. We show how neighborhoods and other principles of local search can be adapted to the specific case of round-robin tournament scheduling problems. We conclude with the presentation of a collection of applications of metaheuristics to problems in sports scheduling.

3.1 Exact and Approximate Optimization Methods

A method for solving an optimization problem is *exact* if it is guaranteed to produce, in finite time, a global optimum for this problem and a proof of its optimality, in case one exists, or otherwise to show that no feasible solution exists. Globally optimal (or, simply, optimal) solutions are often referred to as exact solutions. Exact solution methods for combinatorial optimization problems follow different algorithmic paradigms. We may cite cutting planes, dynamic programming, backtracking, and branch-and-bound (including its variants and extensions, such as branch-and-cut and branch-and-price). Some of these paradigms can be viewed as tree-search procedures: they start from a feasible solution (which corresponds to the root of the tree) and search for the optimal solution by generating and scanning the nodes of a subtree of the solution space (whose nodes often correspond to problem solutions or partial solutions).

Efficient (i.e., running in polynomial time) exact algorithms are unknown (and are unlikely to exist) for a broad class of optimization problems classified as NP-hard. These problems are often referred to as *intractable*. Although the size of the problems that can be solved exactly to optimality has been continuously increasing due to algorithmic and technological developments, there are problems (or problem instances) that are not amenable to being solved by exact methods. Other approaches are needed based on different paradigms to tackle such challenging and large-size optimization problems.

Opposed to exact methods, *approximate methods* or *heuristics* provide feasible solutions that are not necessarily optimal. Approximate methods usually run faster than exact methods. They are capable of handling larger problem instances than exact methods.

Constructive heuristics build a feasible solution from scratch, like the circle method described in Sect. 2.2 to construct a feasible round-robin schedule. In their simplest form, they are often based on low-complexity greedy algorithms. A short introduction to greedy and semi-greedy algorithms is presented in Sect. 3.2.

Local search methods start from a feasible solution and improve it by successive minor modifications until a solution that cannot be further improved is encountered. Although they often provide high-quality solutions whose values are close to those of the optimal solutions, they can become prematurely trapped in low-quality, locally optimal solutions in some situations. Section 3.3 examines solution representations, neighborhoods, search spaces, and local search methods.

Metaheuristics are based on distinct paradigms and offer different mechanisms to escape from locally optimal solutions (as opposed to greedy algorithms or local search methods). They are among the most effective solution strategies for solving challenging combinatorial optimization problems in practice. They often produce much better solutions than those obtained by the simple heuristics and rules they involve. Section 3.4 recalls the principles and templates of some of the most frequently used metaheuristics, which have been instrumental and contributed to most developments and applications in the field.

Section 3.5 introduces the neighborhoods used to implement local search and metaheuristics for scheduling round-robin tournaments. Section 3.6 discusses the connectivity of the solution space induced by these neighborhoods. The related literature in Sect. 3.7 concludes the chapter with a non-exhaustive review of applications of metaheuristics to sports scheduling problems.

3.2 Greedy and Semi-Greedy Algorithms

In a *greedy algorithm*, the solutions are built progressively from scratch. Algorithm 4 provides a template for a greedy algorithm for a minimization problem. The solution S is initialized in line 1. The incremental costs of the elements of the ground set are evaluated in line 2. The loop in lines 3 to 7 is performed while the solution S is not feasible. At each iteration, a new element of the ground set (which preserves the feasibility of the partial solution) with the minimum incremental cost is selected in line 4 and incorporated into the partial solution under construction in line 5. The incremental costs of the ground set elements not belonging to S are updated in line 6, and a new iteration resumes until a feasible solution is obtained. The greedy solution so constructed is returned in line 8.

Algorithm 4: Greedy algorithm for minimization

Output: Greedy feasible solution S

1 $S \leftarrow \emptyset$;
2 Evaluate the incremental cost of each ground set element;
3 **while** *S is not a feasible solution* **do**
4   Select the feasible ground set element $s \in E$ not in solution S with the smallest incremental cost;
5 $S \leftarrow S \cup \{s\}$;
6   Update the incremental cost of each ground set element not in solution S;
7 **end while**
8 Return solution S.

The selection of the following ground set element to be incorporated into the solution is determined by evaluating all candidate elements according to some greedy and easy-to-compute evaluation function. This greedy function usually represents the incremental increase in the cost function due to the incorporation of this element into the partial solution under construction. The greediness criterion establishes that the element with the smallest incremental increase is selected, with ties arbitrarily broken.

The solutions obtained by greedy algorithms are not necessarily optimal. Greedy algorithms are often used to build initial solutions that are needed to be explored by local search or metaheuristics.

Greedy randomized algorithms, or *semi-greedy algorithms*, are based on the same principle of pure greedy algorithms but use randomization at the selection step to build different solutions in different runs of a multi-start procedure. The template of a greedy randomized algorithm for minimization appears in Algorithm 5.

Algorithm 5: Greedy randomized algorithm for minimization

Output: Feasible solution S

1 $S \leftarrow \emptyset$;
2 Evaluate the incremental cost of each ground set element;
3 **while** *S is not a feasible solution* **do**
4   Build a restricted candidate list with the feasible ground set elements not in solution S with the smallest incremental costs;
5   Select an element s from the restricted candidate list at random;
6 $S \leftarrow S \cup \{s\}$;
7   Update the incremental cost of each ground set element not in solution S;
8 **end while**
9 Return solution S.

Compared with the previous algorithm, line 4 of Algorithm 5 builds a restricted candidate list formed by the ground set elements not in the partial solution S that have the smallest incremental costs. In line 5, an element s is randomly selected from the restricted candidate list.

3.3 Local Search

Local search methods start from any feasible solution and visit other (feasible or infeasible) solutions until a feasible solution that cannot be further improved is found. Local improvements are evaluated concerning neighbor solutions that can be obtained by slight modifications applied to the current solution. We first discuss representations of the solutions to combinatorial optimization problems. Next, we define solution neighborhoods, moves between neighbor solutions, and the search space. We present the templates for implementing two fundamental local search strategies: first-improving and best-improving. We conclude this section by presenting the template of variable neighborhood descent, a local search strategy that explores multiple neighborhoods.

3.3.1 Solution Representation, Neighborhoods, and Search Space

A feasible solution satisfies all constraints of the problem at hand. We recall that F is the set of feasible solutions for our combinatorial optimization problem. We denote by $\hat{F} = 2^E$ the set formed by all subsets of the ground set, which includes all feasible and infeasible solutions.

Alternative representations can exist for the solutions of a combinatorial optimization problem. The choice of one over another can lead to more manageable implementations or faster algorithms. We find the following schemes among the most frequently used solution representation techniques:

- *0-1 incidence vector*: This representation is typically used whenever the ground set is partitioned into two subsets, one corresponding to the elements that belong to the solution, while the others do not. Each position of the incidence vector denotes whether or not the corresponding element of the ground set belongs to this solution.
- *Generalized incidence vector*: This representation is used whenever the ground set has to be partitioned into many subsets with different interpretations. Each position of the incidence vector denotes the subset of the partition to which the corresponding element of the ground set belongs.
- *Permutation*: This representation typically applies to scheduling problems in which one is interested in establishing an optimal order for executing some tasks. Each position of the incidence vector denotes the order in which the corresponding element of the ground set will be selected or executed.

A *neighborhood* of a solution $S \in F$ can be defined as any subset of F. More formally, a neighborhood is a mapping that associates each feasible solution $S \in F$ with a subset $N(S) = \{S_1, \ldots, S_p\} \subseteq F$ of feasible solutions.

Each solution $S' \in N(S)$ can be reached from S by an operator called a *move*. Typically, two neighboring solutions S and $S' \in N(S)$ differ only by a few elements of the ground set, and a move from S to S' consists simply in modifying one or more elements in S. Usually, $S \in N(S')$ if only if $S' \in N(S)$.

The *search space* (or *solution space* or *search graph*) of a given problem for some neighborhood N may be seen as a graph $\mathscr{G} = (F, M)$ whose node-set corresponds to its set F of feasible solutions. Its edge-set M is such that there is an edge $(S, S') \in M$ between two solutions $S, S' \in F$ if and only if $S' \in N(S)$ and $S \in N(S')$. An extended search graph may be similarly defined, encompassing not only the set of feasible solutions F but, instead, the whole set $\hat{F} = 2^E$ formed by all subsets of elements of the ground set (feasible or infeasible).

Different neighborhoods can be defined and imposed on the same set of feasible solutions. We observe that some pairs of solutions may be closer in one neighborhood than in another. Furthermore, the search space may be disconnected in some cases. This situation can lead to implementation difficulties and can even make it impossible for the search procedure to find reasonable solutions located in the part of the search space graph that cannot be reached from the initial solution.

This discussion shows that different neighborhoods can be defined and used to implement a local search method. The larger the neighborhood, the denser the search space and the shorter the paths connecting any two solutions. However, the use of large neighborhoods requires the evaluation of more neighboring solutions, leading to more significant computation times during the investigation of the current solution.

It is important to emphasize that the search space does not need to be formed exclusively by feasible solutions in F. It can be enlarged to contain also infeasible solutions. The definitions presented in this section remain the same when the search space is defined over $\hat{F}$ and not only over F. In this situation, a path in the search space can visit feasible and infeasible solutions, but, in any case, it must terminate in a feasible solution. Working with more complex search space graphs, including infeasible solutions, can be essential in some cases to ensure connectivity between any pair of feasible solutions.

Finding an appropriate neighborhood and the best way to explore it is a crucial step toward implementing effective and efficient local search methods.

3.3.2 Local Search Strategies

We have shown that the search space is a graph whose vertices correspond to feasible (and, possibly, infeasible) solutions connected by edges associated with moves to neighbor solutions. A path in the search space consists of a sequence of feasible (and infeasible) solutions in which any two consecutive solutions are neighbors.

Given any instance of a minimization problem, we seek a *globally optimal solution* $S^* \in F$ such that $f(S^*) \leq f(S)$, for all $S \in F$.

A solution S^+ is said to be a *local optimum* for a minimization problem for the neighborhood N if and only if $f(S^+) \leq f(S), \ \forall S \in N(S^+)$. We notice that a global optimum is also locally optimal for any neighborhood, while a local optimum is not necessarily a global optimum.

Local search methods can be viewed as a traversal of the search space starting from any given initial solution and stopping whenever some optimality condition is met. In most cases, a local search procedure is designed to stop after a locally optimal solution is obtained. A local search method's effectiveness and efficiency depend on several factors, such as the starting solution, the neighborhood structure, and the objective function to be optimized. The main components or phases of a local search method are:

Phase 3.1 – ***Start:*** *Construction of the initial solution, from where the search starts. Constructive methods that may be applied to build an initial solution have already been introduced in Sect. 3.2.*

Phase 3.2 – *Neighborhood search:* *Application of a subordinate heuristic or search strategy to find an improving solution in the neighborhood of the current solution. Neighborhood search strategies will be discussed in this section.*

Phase 3.3 – *Stop:* *Interruption of the search by a stopping criterion, which in most cases consists of identifying a locally optimal solution. This chapter will consider stopping criteria for the neighborhood search at different points.*

In what follows, we discuss the first-improving and best-improving strategies for implementing the neighborhood search.

At any iteration of an *iterative improvement* or *first-improving* neighborhood search strategy, the algorithm moves from the current solution to any neighbor with a better (i.e., improving) value for the objective function. The new solution is the first improving solution identified along the neighborhood search. The pseudo-code in Algorithm 6 describes a local search procedure based on the first-improving strategy for a minimization problem. The search starts from a given initial solution S. A flag initialized in line 1 indicates whether or not an improving solution was found. The loop in lines 2 to 10 executes until replacing the current solution with a better neighbor becomes impossible. The flag is reset to .FALSE. in line 3 at the beginning of a new iteration. The loop in lines 4 to 9 visits every neighbor $S' \in N(S)$ of the current solution S until an improving solution is found. If the test in line 5 detects that the neighbor S' is better than the current solution S, then the latter is updated in line 6 Furthermore, the flag is reset to .TRUE. in line 7, indicating that a better solution was found, and a new iteration resumes. The algorithm returns the locally optimal solution S in line 11.

Algorithm 6: First-improving local search for minimization

```
   Input: Initial solution S
   Output: Locally optimal solution S
 1 improvement ← .TRUE.;
 2 while improvement = .TRUE. do
 3     improvement ← .FALSE.;
 4     for every S' ∈ N(S) while improvement = .FALSE. do
 5         if f(S') < f(S) then
 6             S ← S';
 7             improvement ← .TRUE.;
 8         end if
 9     end for
10 end while
11 Return S.
```

In the case of a *best-improving* local search strategy, at any iteration, the algorithm moves from the current solution to its best neighbor whenever the latter improves the former. The pseudo-code in Algorithm 7 describes a local search procedure based on the best-improving strategy for a minimization problem. Once again, the search starts from any given initial solution S. A flag initialized in line 1 indicates whether or not an improving solution was found. The loop in lines 2 to 15 executes until replacing the current solution with a better neighbor becomes impossible. The flag is reset to .FALSE. in line 3 at the beginning of a new iteration. Variable f_{best} that stores the best objective function value over all neighbors of the current solution S is set to a large value in line 4. The loop in lines 5 to 10 visits every neighbor $S' \in N(S)$ of the current solution S. If the test in line 6 detects that the neighbor S' is better than the current best neighbor, then the latter is replaced with S' in line 7, and the best objective function value f_{best} in the neighborhood is updated in line 8. In line 11, we compare the current solution S with its best neighbor S_{best}. If f_{best} is less than $f(S)$, then the current solution is updated in line 12, the flag is reset to .TRUE. in line 13, indicating that a better solution was found, and a new iteration resumes. The algorithm returns the local optimum S in line 16.

Algorithm 7: Best-improving local search for minimization

```
   Input: Initial solution S
   Output: Locally optimal solution S
1  improvement ← .TRUE.;
2  while improvement = .TRUE. do
3      improvement ← .FALSE.;
4      f_best ← ∞;
5      for every S' ∈ N(S) do
6          if f(S') < f_best then
7              S_best ← S';
8              f_best ← f(S');
9          end if
10     end for
11     if f_best < f(S) then
12         S ← S_best;
13         improvement ← .TRUE.;
14     end if
15 end while
16 Return S.
```

We observe that, independently of the starting solution and the neighborhood search strategy, the local search always stops at a local optimum. The complexity of each neighborhood search iteration depends mainly on two factors. First, it depends on the number of neighbors of each visited solution. Second, on the efficiency of the computation of the cost function value for each neighbor. Efficient implementations of the neighborhood search usually compute the cost of each neighbor S' by updating the cost of the current solution S, instead of calculating it from scratch and avoiding repetitive and unnecessary calculations.

Other ingenious implementation tricks, such as the use of candidate lists in best-improving strategies (to reduce the number of moves evaluated in each neighborhood search) or circular search in first-improving strategies (to avoid the reevaluation of non-improving moves already evaluated in the previous neighborhood search) can further improve the efficiency and the effectiveness of local search methods. The use of multiple neighborhoods coordinated with a variable neighborhood descent strategy (to escape from locally optimal solutions by jumping to more complex, higher-order neighborhoods) is discussed in the next section.

3.3.3 *Variable Neighborhood Descent*

A local optimum for a neighborhood is not necessarily optimal for others. The local search procedure can perform neighborhood changes along the descent toward a local optimum. They are crucial and instrumental in some cases. Exploring different neighborhoods in an appropriate order can save a significant amount of computation time.

Small neighborhoods, or those whose elements can be evaluated quickly, may be explored first. The search moves to progressively broader or more complex neighborhoods as locally optimal solutions are found within the lower-order neighborhoods. This local search strategy is called *variable neighborhood descent* (VND), and its main steps are presented in the pseudo-code of Algorithm 8 for a minimization problem, where $N_k(S)$ denotes the set of solutions in the k-th neighborhood of S, for $k = 1, \ldots, k_{max}$.

Algorithm 8: Variable neighborhood descent for minimization

```
Input: Initial solution S
Output: Locally optimal solution S
improvement ← .TRUE.;
while improvement = .TRUE. do
    improvement ← .FALSE.;
    k ← 1;
    while k ≤ k_max do
        Apply a local search strategy to solution S using neighborhood N_k to obtain a local
        optimum S';
        if f(S') < f(S) then
            S ← S';
            k ← 1;
            improvement ← .TRUE.;
        else
            k ← k + 1;
        end if
    end while
end while
Return S.
```

At any iteration of a VND local search strategy, the algorithm moves from the current solution S to a local optimum using the current neighborhood. The search starts from a given initial solution S. A flag that indicates when an improving solution is found is set to .TRUE. in line 1. The loop in lines 2 to 15 is performed until replacing the current solution with a better neighbor becomes impossible. The flag is reset to .FALSE. in line 3 at the beginning of a new iteration. The index of the current neighborhood is set initially to 1 in line 4. The loop in lines 5 to 14 consecutively investigates all neighborhoods. Line 6 applies local search to the current solution S and returns a local optimum S' concerning the neighborhood N_k. The test in line 7 checks if S' improves the current solution S. Then, S is replaced by S' in line 8, the neighborhood counter is reset to 1 in line 9, and the flag is reset to .TRUE. in line 10, indicating that the search found a better solution. Otherwise, the neighborhood counter is increased by one in line 12, indicating that the local search will investigate a higher-order neighborhood next. The algorithm returns the local optimum S concerning all neighborhoods in line 16.

3.4 Metaheuristics and Templates

In this section, we review the basic principles and ideas involved in the design of the main metaheuristics. We give templates for six widely and successfully used metaheuristics: simulated annealing, tabu search, GRASP, VNS, ILS, and genetic algorithms. We also briefly introduce path relinking, an intensification technique that can be combined with virtually any metaheuristic. Sections 3.4.1–3.4.5 provide examples of trajectory-based metaheuristics, in which the initial solution evolves by neighborhood moves (together with perturbations and restarts) until the method stops. Section 3.4.6 presents an introduction to genetic algorithms, which are a typical example of a population-based metaheuristic: a population of solutions evolves along the iterations (named generations in this case) by different types of operations (typically, crossover, mutation, and selection), and it is not possible to trace the individual trajectory of each specific solution.

3.4.1 Simulated Annealing

The principles and basic ideas from simulated annealing come from an analogy with the physical annealing process. In physics, annealing is a thermal process for obtaining the low-energy states of a solid in a heat bath. The principle of simulated annealing lies in the analogy between the annealing process and the combinatorial optimization problem. Solutions to the combinatorial problem are equivalent to states of the physical process, while the cost of each solution corresponds to the state's energy. Neighbor solutions correspond to states generated by a perturbation method.

Algorithm 9 presents the pseudo-code of a simulated annealing algorithm for minimization. The algorithm starts from an initial solution S_0 and computes the initial temperature T_0. The current and the best known solutions are initialized in

Algorithm 9: Template of a simulated annealing algorithm for minimization

Input: Initial solution S_0
Output: Improved solution S^*
1 $S, S^* \leftarrow S_0$;
2 Compute the initial temperature T_0 and set $T \leftarrow T_0$;
3 **while** *stopping criterion not reached* **do**
4   **while** *thermal equilibrium not reached* **do**
5   Obtain a neighbor solution $S' \in N(S)$ at random;
6         $\Delta E \leftarrow f(S') - f(S)$;
7   **if** $\Delta E < 0$ **then**
8             $S \leftarrow S'$;
9   **else**
10   **if** $e^{-\Delta E/(k_B T)} > random[0, 1)$ **then**
11                 $S \leftarrow S'$;
12   **end if**
13   **end if**
14   **if** $f(S) < f(S^*)$ **then**
15             $S^* \leftarrow S$;
16   **end if**
17   **end while**
18   Decrease T according with the annealing schedule;
19 **end while**
20 Return S^*.

line 1 with S_0. The initial temperature T_0 is set in line 2. Iterations are performed in the outer loop in lines 3 to 19 until a specific stopping criterion is satisfied. In most implementations, the stopping criterion brings the temperature to a very small value, usually close to zero. For each temperature T, the inner loop in lines 4 to 17 is performed until a thermal equilibrium is reached. The latter is often implemented by a fixed number of iterations, which depends on the temperature. A solution S' in the neighborhood of the current solution S is randomly generated in line 5. The variation $\Delta E = f(S') - f(S)$ in the objective value is computed in line 6. If line 7 determines that the new solution S' improves the current solution S (i.e., $\Delta E < 0$), then S' replaces S and becomes the new current solution in line 8. The same happens with probability $e^{-\Delta E/(k_B T)}$ in lines 10–11 in case the new solution is not better than the current, where k_B is the Boltzmann constant. Next, the current solution S is compared with the best known solution S^* in line 14, and the latter is updated in line 15 if the new current solution improves upon it. The temperature T is reduced according to an annealing schedule in line 18 once thermal equilibrium is reached. In most implementations, the temperature is geometrically reduced by a factor smaller than one (but close to one to avoid premature convergence). The best solution found S^* is returned by the algorithm in line 20.

The algorithm starts with a high temperature T, allowing even bad solutions to replace the incumbent. As the algorithm proceeds, the temperature T decreases, making it harder to accept low-quality solutions. No further deterioration is accepted when the temperature T approaches zero. Convergence relies on several implementation choices, such as the initial value T_0 of the temperature, the annealing schedule to reduce the temperature, the stopping criterion, and the conditions for reaching the thermal equilibrium at each temperature.

The main advantage of simulated annealing is its simplicity of implementation because there is no local search to be performed. However, its convergence speed and solution quality depend on several parameter values and implementation choices that are not easy to tune. Furthermore, the algorithm is memoryless and does not use cost information about the entire neighborhood.

3.4.2 Tabu Search

Tabu search is a dynamic neighborhood method that uses memory to drive the search by escaping from local optima and avoiding cycling. Contrarily to memoryless heuristics such as simulated annealing, and to methods that use rigid memory structures such as branch-and-bound, tabu search makes use of flexible and adaptive memory designs.

For any solution S, in the case of tabu search, the neighborhood $N(S)$ is not a static set and can be modified according to the history of the search. At each local search iteration, tabu search looks for the neighbor solution that best improves the objective function (or, if none exists, for the one that least deteriorates the current solution). However, some neighbors in $N(S)$ are forbidden and discarded. The set of forbidden (or *tabu*) neighbors is stored in a short-term memory, formed by the most recently visited solutions, to prevent the search from returning to previously visited solutions. Contrary to simple local search procedures, tabu search chooses the move that least deteriorates the objective function value whenever there are no improving moves.

Algorithm 10: Template of a tabu search algorithm for minimization

Input: Initial solution S_0
Output: Locally optimal solution S^*

$S, S^* \leftarrow S_0$;
Initialize the tabu list $T \leftarrow \emptyset$;
while *stopping criterion not reached* **do**
  Obtain the best neighbor S' of the current solution S in $N(S) \setminus T$;
  $S \leftarrow S'$;
  if $f(S) < f(S^*)$ **then**
    $S^* \leftarrow S$;
  end if
  Update the tabu list T;
end while
Return S^*.

The basic steps of a simple tabu search heuristic for minimization, based exclusively on short-term memory, are described in the pseudo-code of Algorithm 10. The algorithm starts from an initial solution S_0. The current and the best solutions are initialized with S_0 in line 1, The short-term memory T is initialized in line 2. Iterations are performed in the loop in lines 3 to 10 until a specific stopping criterion is reached. In most implementations, it corresponds to an upper limit on the number of consecutive moves without improvement in the best solution value. Each iteration starts in line 4 with the search for the best non-forbidden neighbor S' of the current solution S, which becomes the current solution in line 5. Next, the best known solution S^* is updated in line 7 if the new current solution S improves it. The next step concludes the iteration by updating the tabu list in line 9, and a new iteration resumes. The best local optimum found S^* is returned by the algorithm in line 11.

The tabu search scheme described above makes use of some essential features. It can be extended by incorporating more sophisticated strategies: aspiration criteria to override the tabu status of unvisited solutions, medium-term and long-term memories to implement intensification and diversification procedures, candidate list strategies to speed up the search, and hashing tables to accelerate and filter the search. The tabu list can be implemented using different strategies, such as storing moves instead of solutions or forbidding only the reversal of non-improving moves but not improving moves. These comprehensive strategies lead to much more powerful and effective implementations in terms of the quality of the solutions they obtain.

Tabu search is undoubtedly among the most effective approaches for solving hard combinatorial optimization problems. It has been successfully applied to many problems in numerous domains. However, implementations of tabu search often involve setting many parameters, which have to be appropriately tuned for achieving good performance in practice.

3.4.3 Greedy Randomized Adaptive Search Procedures

A greedy randomized adaptive search procedure (GRASP) is a metaheuristic resulting from the hybridization of a semi-greedy algorithm with a local search method embedded in a multistart framework. The method consists of multiple applications of local search (see Sect. 3.3), each starting from a solution generated by a semi-greedy algorithm (see Sect. 3.2) implementing the construction procedure.

The pseudo-code of Algorithm 11 illustrates the main blocks of a GRASP procedure for minimization. Line 1 initializes the best solution value. Iterations are performed in the loop in lines 2 to 9 until a specific stopping criterion is reached. The maximum number of iterations is typically used as the stopping criterion. Line 4 builds a feasible solution S using a semi-greedy construction. Local search is applied to S, and a local minimum S' is obtained. The best local optimum S^* is updated in line 6, and the best solution value is updated in line 7. Line 10 returns the best local optimum S^* obtained over all iterations as the solution provided by the algorithm.

Algorithm 11: Template of a GRASP algorithm for minimization

Input: Initial seed
Output: Locally optimal solution S^*

$f^* \leftarrow \infty$;
while *stopping criterion not reached* **do**
  Build a feasible solution S using a semi-greedy algorithm and the current seed;
  Obtain a local optimum S' by applying local search to solution S;
  if $f(S') < f^*$ **then**
    $S^* \leftarrow S'$;
    $f^* \leftarrow f(S^*)$;
  end if
end while
Return S^*.

An especially appealing characteristic of GRASP is its ease of implementation. Few parameters must be set and tuned; therefore, development can focus on implementing efficient data structures to ensure fast computations in all iterations. Basic implementations of GRASP rely almost exclusively on two parameters. The first is the stopping criterion, often defined by the maximum number of iterations (or by the maximum number of iterations without improvement in the best solution found). The second is a parameter used to limit the size of the restricted candidate list within the greedy randomized algorithm used in the construction phase. Despite its simplicity and ease of implementation, GRASP is a very effective metaheuristic and produces the best known solutions for many problems.

GRASP, as initially proposed, is a memoryless procedure in which each iteration does not use information gathered in previous iterations. In other words, all its iterations are independent. Path relinking enhances the basic GRASP significantly, adding a long-term memory mechanism to GRASP heuristics. GRASP with path relinking implements long-term memory using an elite set of diverse and high-quality solutions previously found during the search. In its most basic implementation, path relinking is applied at each iteration between the solution found at the end of the local search phase and any randomly selected solution from the elite set. The solution resulting from path relinking is a candidate for inclusion in the elite set. The hybridization of GRASP with path relinking can improve solution time and quality significantly. The pseudo-code in Algorithm 12 illustrates the main steps of a GRASP heuristic with path relinking for minimization. It implements a memory-based intensification strategy.

Algorithm 12: Template of a GRASP with path relinking for minimization

Input: Initial seed
Output: Locally optimal solution S^*

$f^* \leftarrow \infty$;
Initialize the pool of elite solutions: $P \leftarrow \emptyset$;
while *stopping criterion not reached* **do**
  Build a feasible solution S using a semi-greedy algorithm and the current seed;
  Obtain a local optimum S' by applying local search to solution S;
  if $f(S') < f^*$ **then**
    $S^* \leftarrow S'$;
    $f^* \leftarrow f(S^*)$;
  end if
  if $P \neq \emptyset$ **then**
    Select a solution $S'' \in P$ at random;
    Obtain a solution $\hat{S}$ by applying path relinking to S' and S'';
    Obtain a new local optimum S' by applying local search to solution $\hat{S}$;
    if $f(S') < f^*$ **then**
      $S^* \leftarrow S'$;
      $f^* \leftarrow f(S^*)$;
    end if
  end if
  Update the pool of elite solutions P with the local optimum S';
end while
Return S^*.

3.4.4 Variable Neighborhood Search

Variable neighborhood search (VNS) is based on a systematic change of neighborhood. It makes use of a finite set of k_{max} pre-selected neighborhood structures identified as $N_1, N_2, \ldots, N_{k_{max}}$ that may be induced by one or more metric functions introduced in the solution space. Let $N_k(S)$ denote the set of solutions in the k-th neighborhood of S.

Algorithm 13 summarizes the main steps of the pseudo-code of a VNS heuristic for minimization. The algorithm starts from an initial solution S initialized with S_0 in line 1. Iterations of the outer loop are performed in lines 2 to 14 until some specific stopping criterion is reached. Typically, this can be the maximum number of times the highest order neighborhood $N_{k_{max}}$ is attained. At the beginning of each iteration, the current neighborhood is set to $k = 1$ in line 3, i.e., the search starts from the lowest order neighborhood $N_1(S)$ of the current solution S. Neighborhoods N_1 to $N_{k_{max}}$ of the current solution are progressively scanned within the inner loop in lines 4 to 13. The initial step of each iteration of this inner loop is a random perturbation applied to generate a solution S' in the current neighborhood $N_k(S)$ in line 5. Local search is applied to S', and a local minimum S'' is obtained in line 6. If the local minimum S'' improves the current solution S, then the latter is updated with S'' in line 8, and the current neighborhood is reset to $k = 1$ in line 9 (i.e., the search will resume from neighborhood $N_1(S)$). Otherwise, the current solution S is not modified, but the current neighborhood k is incremented by one in line 11 (i.e., the search will resume from a higher-order neighborhood of S). Line 15 returns the best solution visited S.

Algorithm 13: Template of a VNS algorithm for minimization

```
   Input: Initial solution S0, initial seed
   Output: Locally optimal solution S
1  S ← S0;
2  while stopping criterion not reached do
3      k ← 1;
4      while k ≤ k_max do
5          Obtain a neighbor solution S' ∈ N_k(S) at random;
6          Obtain a local optimum S'' by applying local search to solution S';
7          if f(S'') < f(S) then
8              S ← S'';
9              k ← 1;
10         else
11             k ← k + 1;
12         end if
13     end while
14 end while
15 Return S.
```

VNS is also a memoryless heuristic. It is easy to implement and relies on a few parameters: the stopping criterion and the number k_{max} of neighborhoods (and their definitions).

3.4.5 *Iterated Local Search*

Iterated local search (ILS) explores the solution space using a walk that moves from a locally optimal solution to a nearby one without the restriction of using exclusively moves in the neighborhood. Algorithm 14 gives the pseudo-code of ILS. It first applies local search to the initial solution in line 1. The loop in lines 2 to 8 is performed until some stopping criterion is reached. Each iteration starts by applying a random perturbation to the current locally optimal solution S, which results in an intermediate solution S' in line 3. Local search is again applied to S' in line 4, resulting in a new local optimum S''. Line 5 checks if the new solution S'' meets some acceptance criterion, in which case it becomes the next solution of the solution trajectory in line 6. Otherwise, the algorithm remains in the same locally optimal solution S, and a new iteration resumes. A maximum number of perturbations is typically used as the stopping criterion.

Algorithm 14: Template of an ILS algorithm for minimization

Input: Initial solution S_0
Output: Locally optimal solution S

Obtain a local optimum S by applying local search to solution S_0;
while *stopping criterion not reached* **do**
    Obtain a new solution S' by applying a random perturbation to the current solution S;
    Obtain a local optimum S'' by applying local search to solution S';
    if S'' *satisfies an acceptance criterion* **then**
        $S \leftarrow S''$;
    end if
end while
Return S.

3.4.6 Genetic Algorithms

A genetic algorithm (GA) is a population-based method that uses selection, recombination, and mutation operators to generate new solutions in the search space, imitating the process of natural selection and evolution of species. Solutions are evaluated in terms of their fitness, which most often corresponds to the value of the objective function itself.

Contrarily to the other metaheuristics reviewed in Sects. 3.4.1–3.4.5, genetic algorithms explore a population of solutions that evolve along some generations. There is not a single trajectory emanating from a unique initial solution. Given a population P formed by a fixed number p of feasible solutions to the problem at hand at some generation, the following operations are involved in standard implementations of genetic algorithms:

- `EvaluateFitness`: the fitness of every solution in the population is evaluated.
- `SelectMates`: a subset of solutions is selected for mating, either randomly or using fitness information to privilege promising solutions that could generate well-fitted offspring.
- `CombineParents`: pairs of solutions selected by the operator `SelectMates` are combined, and new solutions (offspring) are generated (this is the *crossover* operator).
- `ApplyMutation`: a few randomly selected solutions go through small random modifications in their structure (this is the *mutation* operator).
- `SelectNewPopulation`: the best-fitted solutions of the population are selected and passed to the new generation (this is the *selection* operator).

The pseudo-code of the basic steps of a genetic algorithm for minimization appears in Algorithm 15. The generation counter is initialized in line 1. The initial population P (formed by p feasible solutions) is generated in line 2. There are several strategies to generate the initial population P, such as entirely randomly or by using a greedy randomized algorithm. The fitness of each solution of the

Algorithm 15: Template of a genetic algorithm for minimization

Output: Best solution found S^*
1 $k \leftarrow 0$;
2 Generate the initial population P formed by p feasible solutions;
3 `EvaluateFitness`(P);
4 **while** *stopping criterion not reached* **do**
5 $k \leftarrow k + 1$;
6 *Parents* $\leftarrow$ `SelectMates`(P);
7 *Offspring* $\leftarrow$ `CombineParents`(*Parents*);
8 $P' \leftarrow$ `ApplyMutation`(*Offspring* $\cup$ P);
9 `EvaluateFitness`(P');
10 $P \leftarrow$ `SelectNewPopulation`(P');
11 **end while**
12 Let S^* be the best fitted solution in P (i.e., the minimum cost solution in P);
13 Return S^*.

population is evaluated in line 3. The loop in lines 4 to 11 is performed until some stopping criterion is not reached. Most applications of genetic algorithms use the total number of generations or the stabilization of the best solution in the population as the stopping criterion. Each iteration of this loop handles one generation of the population. The generation counter is updated in line 5. The process of obtaining a new generation starts in line 6, selecting a subset *Parents* of solutions from the current population. The selection for mating may be done either randomly or taking into account the fitness functions to extract from the populations the best candidates to generate well-fitted offspring. Solutions from *Parents* are pairwise combined by a crossover operator in line 7, generating a set *Offspring* of new solutions. The crossover operator can be performed at random or may give a higher probability of preserving the characteristics of the best-fitted parent. Some implementations attempt to generate optimal offspring or apply a local search procedure after crossover. A small subset of the solutions in *Offspring* $\cup P$ are submitted to mutations in line 8, from which a new population P' results. The fitness of each solution in P' is evaluated in line 9. Finally, a selection operator is applied, and the p best solutions in population P' are chosen to form the population P for the new generation when a new iteration resumes. Most implementations apply the selection operator to the entire renewed population P', while some apply it only to the newly generated *Offspring* population. In the end, the best solution in the last generation is returned. The algorithm follows an elitist strategy, and the best solution is always preserved from generation to generation.

Initially, genetic algorithms did not incorporate any problem-specific knowledge and relied almost entirely on randomization for the initial population generation, as well as for mate selection, crossover, and mutation. This characteristic has progressively changed with time.

Hybrid genetic algorithms, also called memetic algorithms, use specific knowledge available to solve the problem. Such strategies incorporate greedy randomized algorithms to create the initial population and the application of local search to some elements of the population of each generation. Genetic algorithms hybridized with optimization strategies usually perform much better than basic genetic algorithms.

Genetic algorithms are very appealing and easy to implement. However, many implementation choices for each of the above operators are possible. An exemplary implementation requires an appropriate solution encoding as finite-length data strings and well-tuned strategies for generating the initial population, mate selection, crossover, mutation, and selection.

A significant difficulty with implementing genetic algorithms resides in the crossover operator. This difficulty occurs because, in most cases, the solutions obtained by crossover turn out to be infeasible, therefore, unusable by the algorithm. Although some strategies have been proposed in the literature to handle infeasibilities (such as the use of penalties or repair operators), they are problem specific and of complex generalized implementation in practice, with a limited number of successful cases.

A *biased random-key genetic algorithm* (BRKGA) encodes solutions as a vector of real numbers called *random keys*. A deterministic algorithm, called a *decoder*, takes a vector of random keys as input and associates with it a feasible solution of the combinatorial optimization problem at hand, for which an objective value or fitness function can be computed. Crossover is said to be biased because one parent is always an elite (high-quality) solution, while the other is not. The elite parent still has a higher probability of passing its elements to the new generation. Biased random-key genetic algorithms can be implemented straightforwardly. Crossover is performed in the space of random keys, and the corresponding solutions are built by the decoder a posteriori. Therefore, the resulting solutions are always feasible. BRKGAs are particularly effective for scheduling problems or other combinatorial optimization problems whose solutions may be represented by permutation vectors.

3.5 Neighborhoods for Round-Robin Tournaments

We noticed in Sect. 3.3 that alternative representations might exist for the solutions of a combinatorial optimization problem. We also observed that different neighborhoods could be defined and imposed on the set of feasible solutions. Multiple neighborhoods can be handled by local search methods such as variable neighborhood descent (VND) or metaheuristics such as variable neighborhood search (VNS).

We have seen in Sect. 3.4 that virtually all metaheuristics have local search methods as one of their components. Local search methods embedded in metaheuristics may generate neighbors randomly, as in the case of simulated annealing. They may also explore solution neighborhoods using first-improving or best-improving strategies, as in the cases of tabu search or GRASP. Other more advanced strategies, such as VNS, may combine several neighborhoods.

Solutions for single round-robin tournament scheduling problems will be represented here by oriented ordered 1-factorizations of the complete graph K_{2n} for a league with $2n$ teams. This representation may also be used for problems associated with mirrored double round-robin scheduling tournaments. With this representation, each 1-factor is associated with a different color, which is used to color all edges of this 1-factor.

Different neighborhood structures have been used in the literature of local search procedures for scheduling single round-robin sports tournaments, although sometimes they have been referred to by different names; see Sect. 3.7 for a short historical discussion. We name them as *Home-Away Swap, Team Swap, Partial Team Swap, Round Swap, Partial Round Swap,* and *Generalized Partial Team Swap*. Edge orientations are disregarded for all neighborhoods except Home-Away Swap. These neighborhoods may be extended to the case of double round-robin problems. They are described in the following sections, using the notation and definitions introduced in Chap. 2.

3.5.1 Neighborhood "Home-Away Swap" (HAS)

Moves in this neighborhood operate only on home-away patterns, leaving the timetable unchanged. The operator that defines the neighborhood consists of taking an arc of the oriented 1-factorization and reversing its orientation. In practice, it corresponds to exchanging the venue of the game between the two teams associated with the endpoints of this arc.

3.5.2 Neighborhood "Team Swap" (TS)

Each move in this neighborhood is obtained by swapping two distinct nodes of the graph. After a TS move is applied to two nodes $v \neq w$, for every node $x \neq v$ and $x \neq w$, the arc between x and v gets the color previously used on the arc between x and w, and vice-versa. In practice, it corresponds to exchanging the opponents of the teams associated with nodes v and w in all tournament rounds. Figure 3.1 illustrates the application of a move in neighborhood TS to teams 3 and 4 for a schedule with six teams.

3.5.3 Neighborhood "Round Swap" (RS)

Moves in this neighborhood pick two distinct colors, j and k, and exchange them in all edges colored with any of these colors. In other words, we swap the two 1-factors F_j and F_k associated with the colors j and k in the ordered 1-factorization. In practice, it corresponds to swapping the two rounds j and k in the tournament

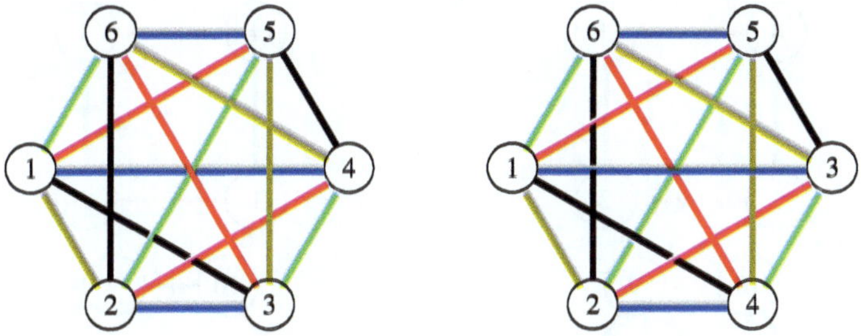

(a) (b)

Fig. 3.1 Graph on the left illustrates a 1-factorization of K_6. On the right, we see the 1-factorization obtained by a team swap move applied to teams (nodes) 3 and 4. (**a**) Current solution. (**b**) Solution after move TS

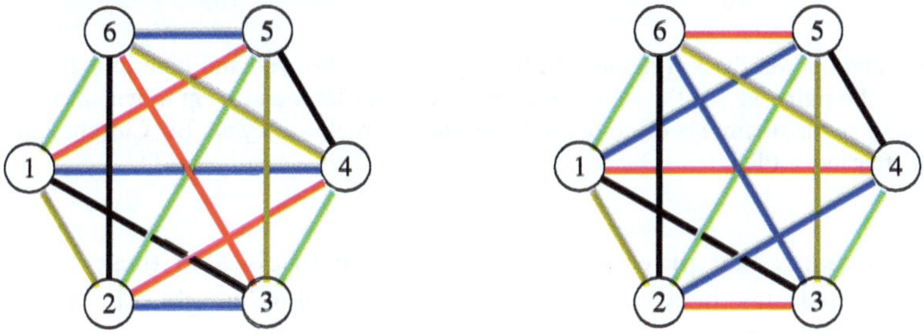

(a) (b)

Fig. 3.2 Graph on the left illustrates a 1-factorization of K_6. On the right, we see the 1-factorization obtained by a round swap move applied to the colors red and blue. (**a**) Current solution. (**b**) Solution after move RS

sequence. Figure 3.2 illustrates the application of a move in neighborhood RS to the colors red and blue for a schedule with six teams.

3.5.4 Neighborhood "Partial Round Swap" (PRS)

For moves in this neighborhood, we select any two distinct colors and consider any cycle C in the subgraph formed by the edges colored with these colors. A neighbor is obtained by exchanging the colors in this cycle. If the edges of this cycle are incident to all vertices of the graph, i.e., if they form a Hamiltonian cycle, then this move is equivalent to a move in the RS neighborhood. Consequently, PRS and PS are equivalent in the case of perfect 1-factorizations; see Sect. 2.1. In practice, it corresponds to picking two rounds in which there is a subset formed by an even

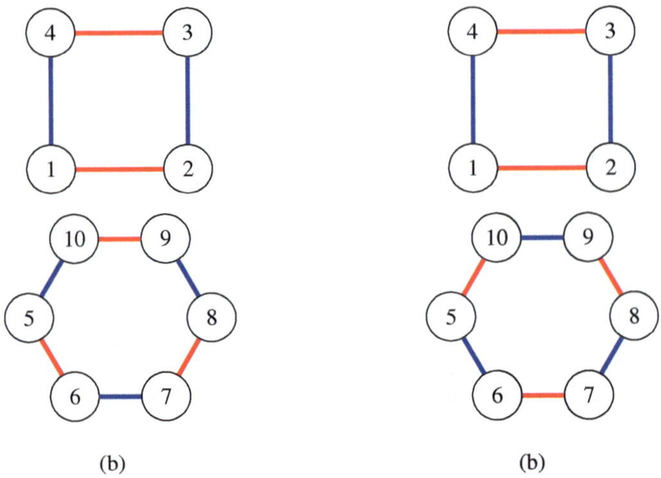

Fig. 3.3 Subgraph on the left is induced by edges with colors red and blue in a 1-factorization of K_{10}. On the right, we see the same subgraph after a partial round swap move has been applied by swapping the assignments of colors red and blue in the lower cycle. (**a**) Current solution. (**b**) Solution after move PRS

number of teams that play exclusively among themselves in these two rounds. Figure 3.3 illustrates the application of a move in the PRS neighborhood.

A move in the neighborhood RS is equivalent to a sequence of moves in the neighborhood PRS.

3.5.5 Neighborhood "Partial Team Swap" (PTS)

A move in this neighborhood takes two nodes v_1 and v_2 and a collection of chains $C = \{C_1, \ldots, C_p\}$ of length two between them. The set $\mathscr{S}$ of colors assigned to the edges of the chains in C adjacent to v_1 should be the same as the set of colors assigned to the edges of the chains in C adjacent to v_2. We choose $\mathscr{S}$ to be minimal, i.e., no proper subset of C satisfies this same condition. The edge-colored subgraph composed by v_1, v_2, and the chains in C is called a *colorful chordless lantern*. Then, the two colors of each chain are exchanged.

We notice that if the number of colors occurring in these chains is always $2n - 2$ for whatever pair of nodes v_1 and v_2 that are picked, then each PTS move is equivalent to a TS move. Such 1-factorizations have been named *L-blocking* in the literature because they block the possibility of obtaining a non-isomorphic 1-factorization by exchanging the colors of a colorful chordless lantern.

This move corresponds to picking the two teams corresponding to nodes v_1 and v_2 and a minimal subset of rounds in which both of them face the same opponents and then exchanging the opponents of v_1 and v_2 in each of these rounds. Figure 3.4

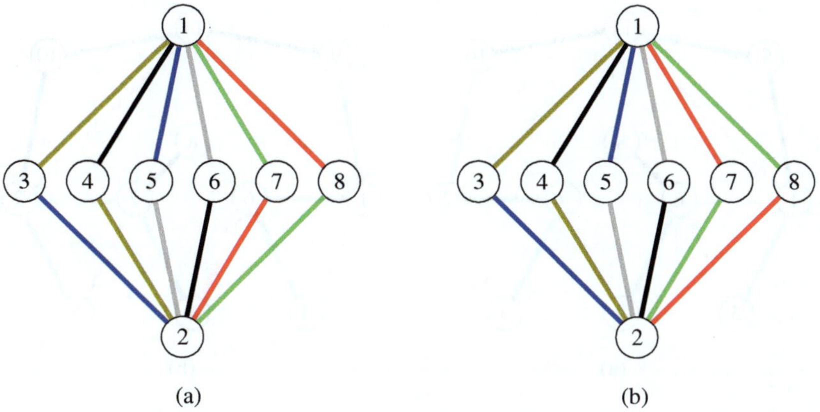

Fig. 3.4 Subgraph on the left is obtained from a 1-factorization of K_8. On the right, we see the same subgraph after a partial team swap move with chains $C_1 =< v_1 = 1, 7, v_2 = 2 >$ and $C_2 =< v_1 = 1, 8, v_2 = 2 >$. (**a**) Current solution. (**b**) Solution after move PTS

illustrates the application of a move in the PTS neighborhood with nodes $v_1 = 1$ and $v_2 = 2$.

A move in the neighborhood TS is equivalent to a sequence of moves in the neighborhood PTS.

3.5.6 Neighborhood "Generalized Partial Team Swap" (GPTS)

This neighborhood involves two vertices v_1 and v_2 and a set of edge-disjoint alternating chains $\{C_1, \dots, C_p\}$ of even length. We observe that GPTS generalizes both PRS and PTS. It generalizes PTS by allowing the chains to be longer than two. To see that it generalizes PRS, take v_1 and v_2 as two arbitrary nodes at even distance in the cycle C, then the latter can be seen as two alternating chains of even length between v_1 and v_2. Figure 3.5 illustrates the application of a move in the GPTS neighborhood with nodes $v_1 = 1$ and $v_2 = 2$.

Due to the size and complex structure of neighborhood GTPS, it might be hard to design algorithms that systematically explore this neighborhood. Efficient strategies for obtaining a set of edge-disjoint alternating chains $\{C_1, \dots, C_p\}$ of even length between v_1 and v_2 need to be designed. Also, efficient techniques to recognize that the same colors are present in v_1 and v_2 would be essential.

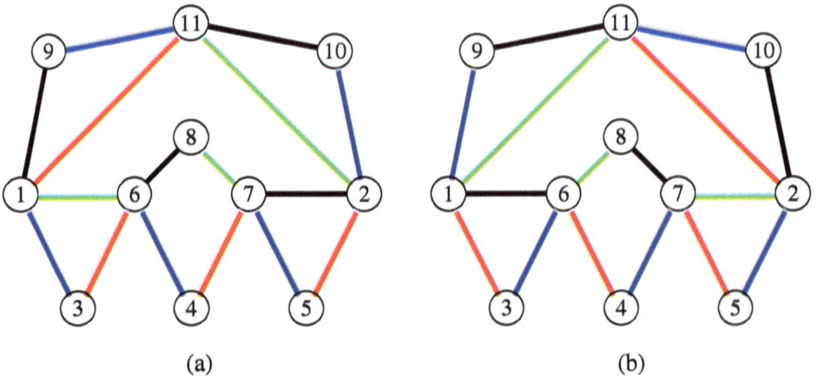

Fig. 3.5 Subgraph on the left is obtained from a 1-factorization of K_{12}. On the right, we see the same subgraph after a generalized partial team swap move involving four alternating chains between nodes $v_1 = 1$ and $v_2 = 2$ (not all nodes of K_{12} appear in the figure). (**a**) Current solution. (**b**) Solution after move GPTS

3.5.7 Ejection and Repair Chains

Neighborhood structures HAS, RS, TS, PRS, and PTS, or some of their variants or restrictions, have been used in virtually all applications of local search heuristics for scheduling round-robin tournaments in the literature, possibly with different names or interpretations.

Single-step moves, such as those described in the previous sections, can be extended to define broader neighborhoods associated with compound moves that can be seen as sequences of simple moves that introduce structural changes in the current solution. Algorithms that incorporate compound moves are called *variable neighborhood methods* since the number of components of a compound move may vary from step to step.

Ejection chains, sometimes called *repair chains*, are variable depth methods that use a sequence of interrelated, consecutive, straightforward moves to create more complex moves. They are designed to induce successive changes following an initial move that entailed infeasibilities in the neighbor solution, until feasibility is recovered. An ejection chain has a variable depth because the number of more straightforward moves needed to recover feasibility may vary from one iteration to another. The sequence of simple consecutive moves leading to a compound move may be long, and evaluating a compound move defining an ejection chain can be very costly regarding the computational effort. Therefore, the full exploration of neighborhoods defined by ejection chains can hardly be done. However, ejection chains may be employed as random or biased perturbations to destroy the structure of a local optimum, followed by the generation of a new solution that still shares part of the original local optimum that produced it.

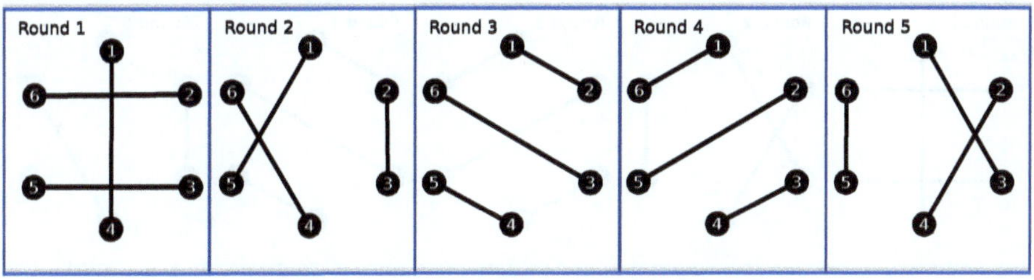

Fig. 3.6 Initial solution corresponding to an ordered 1-factorization of K_6

Some metaheuristics use diversification strategies to drive the search toward unexplored regions of the search space. In particular, iterated local search (ILS) explores perturbation moves to escape from locally optimal solutions, obtaining new, different initial solutions from where the search restarts. Ejection chains are a very attractive alternative to generating perturbation moves in the context of diversification steps in tabu search or iterated local search.

Ejection chains have been used in some heuristics for tournament scheduling problems. We illustrate their application in scheduling a compact single round-robin tournament with $2n$ teams. Since each team plays exactly once with every other team, the tournament takes $2n - 1$ rounds to be completed, and every team plays exactly one game in each round. The tournament is represented by the complete graph K_{2n}, in which each node represents a team, and each edge corresponds to the game between two teams. Figure 3.6 illustrates a feasible schedule for a tournament with six teams, corresponding to an ordered 1-factorization of K_6.

The discussion that follows regards exclusively feasibility issues. The *game rotation* neighborhood, proposed in the literature in the form of an ejection chain, consists in enforcing any given game to be played in any particular round. We denote by $GR(u, v, k)$ the move corresponding to removing the game between teams u and v from the round where it is played in the current solution and rescheduling it in round k. Nevertheless, infeasibilities are created when a game is removed from one round and enforced to be played in another. Appropriate modifications to avoid clashes between teams playing more than once or not playing at all in a given round should be applied. New edges have to be successively removed from one round and reassigned to another until these infeasibilities are eliminated, and a feasible solution is recovered.

The sequence of reassignments of games to new rounds gives rise to an ejection chain, as illustrated in Figs. 3.7, 3.8, and 3.9 for a tournament with six teams. Edges in green and red are ejected from one 1-factor to another at each step. In Fig. 3.7a, the game $(1, 3)$ is selected to be reassigned from round 5 to round 2. In other words, in this example, move $GR(1, 3, 2)$ will be applied to the solution in Fig. 3.6. In each step of Fig. 3.7b, 3.8, and 3.9b, one game is ejected to a new round, and another one is reassigned. Finally, in Fig. 3.9c, the game $(2, 3)$ is assigned to round 5, and a feasible solution is reached, i.e., a new ordered 1-factorization is obtained, with game $(1, 3)$ played in round 2.

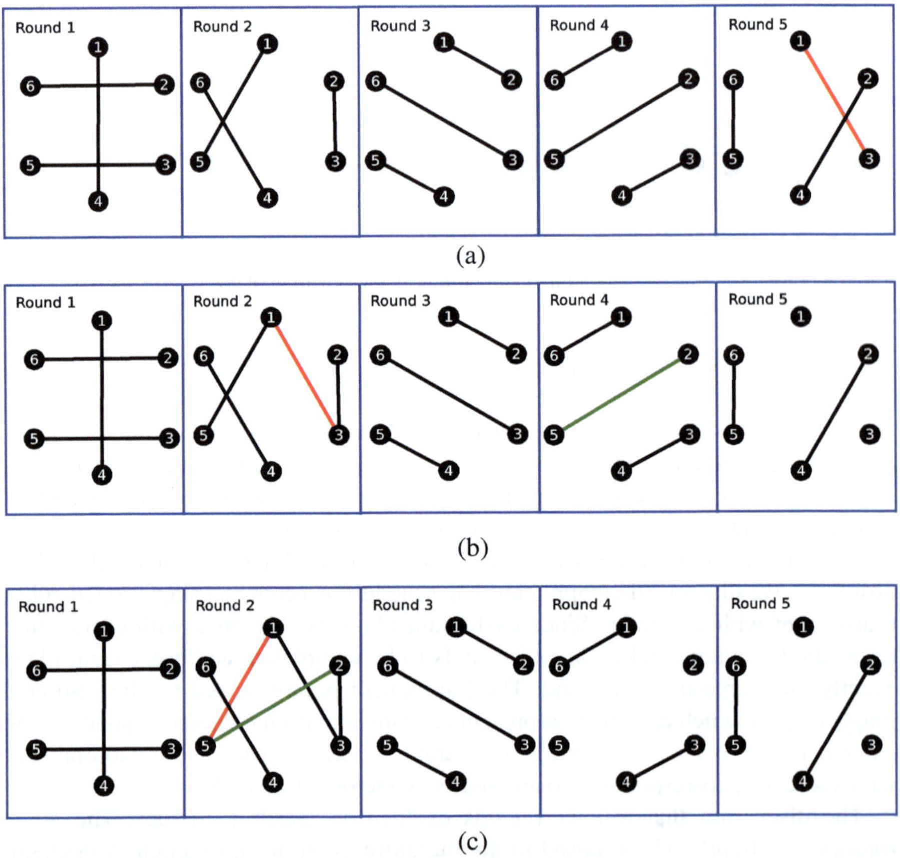

Fig. 3.7 Ejection chain moving game (1, 3) from round 5 to round 2: first three moves. **(a)** Game (1, 3) will be moved from round 5 to round 2. **(b)** Game (1, 3) is now played in round 2, while game (2, 5) will be moved from round 4 to round 2. **(c)** Game (2, 5) is now played in round 2, while game (1, 5) will be moved from round 2 to round 4

We may now observe that the ejection chain illustrated in Figs. 3.6, 3.7, 3.8, and 3.9 is equivalent to the move $PTS(1, 2, \{< 1, 3, 2 >, < 1, 5, 2 >, < 1, 6, 2 >, < 1, 4, 2 >\})$ depicted in Fig. 3.10 and defined by nodes $v_1 = 1$ and $v_2 = 2$, and $C = \{< 1, 3, 2 >, < 1, 5, 2 >, < 1, 6, 2 >, < 1, 4, 2 >\}$. The ejection chain associated with the game rotation move $GR(u, v, k)$ is equivalent to the PTS move involving teams u, the opponent of team v in round k, and a collection of length two chains between u and the opponent of team v in round k in which the color of edge $[u, v]$ is used.

Therefore, although ejection and repair chains have been proposed in the literature for tournament scheduling problems, they can be seen as alternative ways to describe or implement moves in neighborhoods such as PTS and PRS.

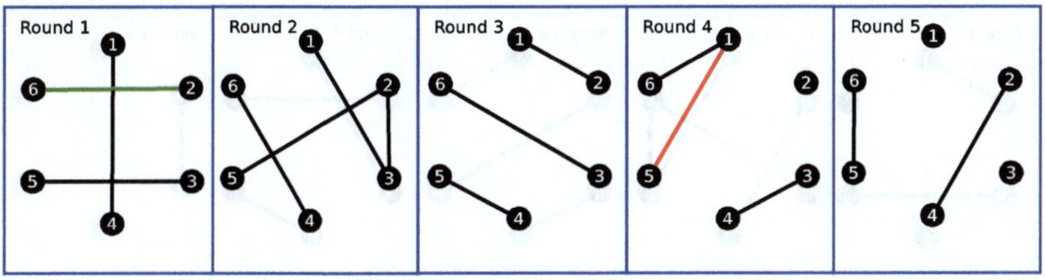

(a)

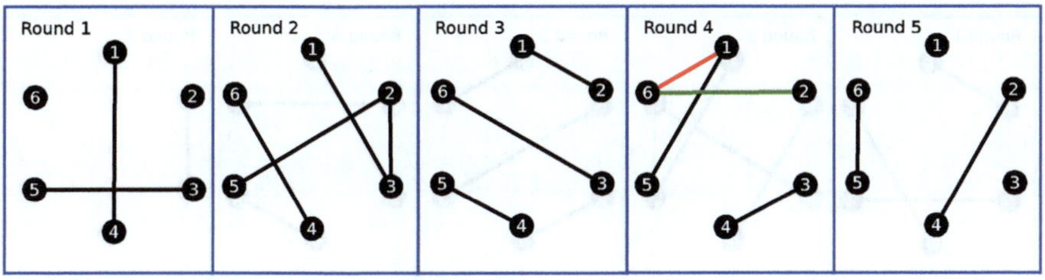

(b)

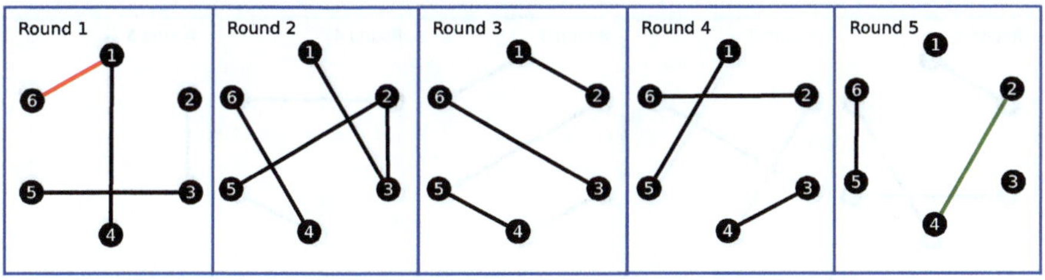

(c)

Fig. 3.8 Ejection chain moving game $(1, 3)$ from round 5 to round 2: intermediate three moves. (**a**) Game $(1, 5)$ is now played in round 4, while game $(2, 6)$ will be moved from round 1 to round 4. (**b**) Game $(2, 6)$ is now played in round 4, while game $(1, 6)$ will be moved from round 4 to round 1. (**c**) Game $(1, 6)$ is now played in round 1, while game $(2, 4)$ will be moved from round 5 to round 1

Ejection chains are often costly in terms of computation time to be used as regular moves in local search methods that should fully explore their neighborhoods. However, they are very effective in generating perturbations and in search diversification.

3.6 Connectivity of the Tournament Space

Let $\mathscr{G} = (F, M)$ be the search space graph associated with some round-robin tournament scheduling problem. The node-set F is formed by all ordered oriented 1-factorizations of K_{2n}, where $2n$ is the number of teams. Given a neighborhood

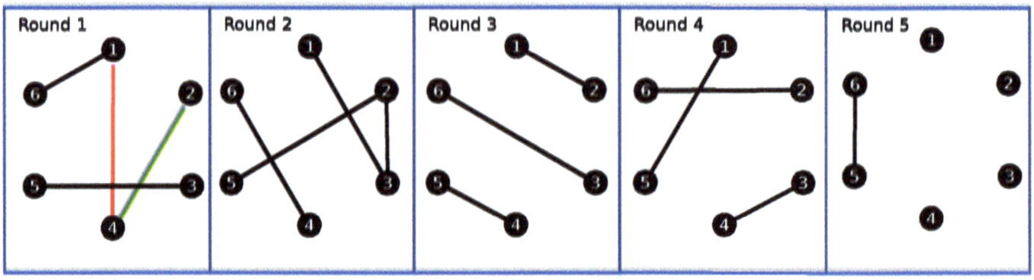

(a)

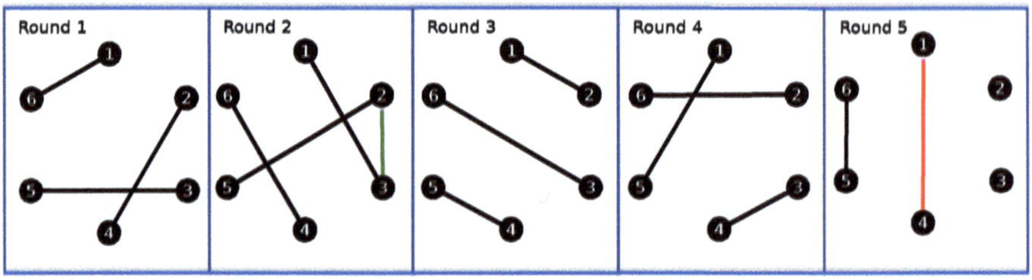

(b)

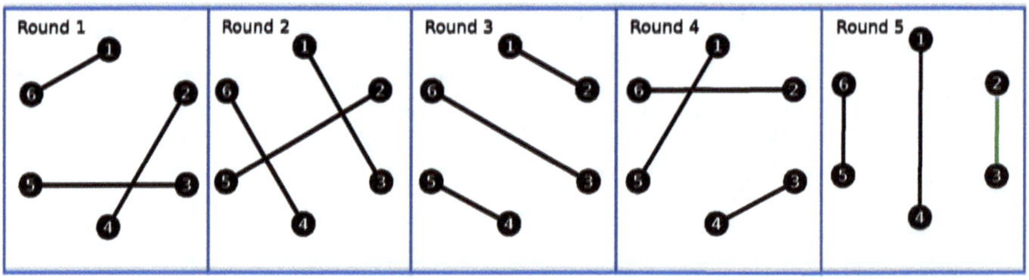

(c)

Fig. 3.9 Ejection chain moving game $(1, 3)$ from round 5 to round 2: final three moves. (**a**) Game $(2, 4)$ is now played in round 1, while game $(1, 4)$ will be moved from round 1 to round 5. (**b**) Game $(1, 4)$ is now played in round 5, while game $(2, 3)$ will be moved from round 2 to round 5. (**c**) Game $(2, 3)$ is now played in round 5 and a new feasible schedule is finally obtained

structure $\mathscr{N}$, two nodes $u, v \in F$ are adjacent, i.e., $[u, v] \in M$, if and only if $u \in \mathscr{N}(v)$ and $v \in \mathscr{N}(u)$. The question arises to determine whether $\mathscr{G}$ is connected or not. If not, starting from some initial solution and applying a local search method, it might be impossible to reach some solutions because they are in different connected components of $\mathscr{G}$.

The size of the search space graph $\mathscr{G} = (F, M)$ of single round-robin tournament scheduling problems is huge even when compared with those of other classical combinatorial optimization problems, such as the traveling salesman problem. There is no known formula for the number of different 1-factorizations of K_{2n}. The number of non-isomorphic 1-factorizations of K_{2n} increases rapidly with $2n$. There are 526,915,620 non-isomorphic 1-factorizations of K_{12} and

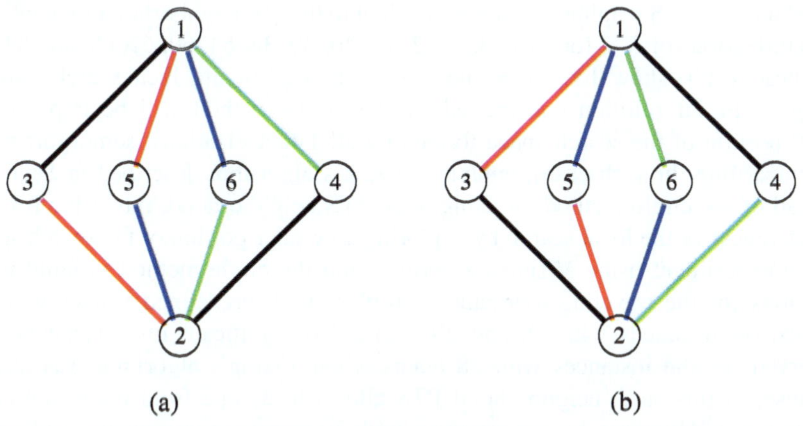

Fig. 3.10 Partial team swap move equivalent to the game rotation ejection chain. (**a**) Current solution. (**b**) Solution after move PTS (ejection chain GR)

1,132,835,421,602,062,347 of K_{14}. The number of single round-robin schedules is even much higher since for each of these non-isomorphic 1-factorizations, there are different round permutations, different home-away for each game, and different teams' assignments to nodes.

In a perfect 1-factorization, the subgraph containing all vertices and edges belonging to the union of any two 1-factors is a Hamiltonian cycle. A perfect 1-factorization may not be desirable as the initial solution for local search algorithms for tournament scheduling problems because every move in the PRS neighborhood has the same effect as a move in the RS neighborhood, generating colorings that are isomorphic to the original one.

The circle method, introduced in Sect. 2.2, is a simple strategy to obtain initial solutions for local search heuristics for single round-robin tournament scheduling problems. However, it generates only one type of 1-factorizations for K_{2n}, the canonical 1-factorization. It generates perfect 1-factorizations for every complete graph K_{2n} for which $2n = p + 1$, where p is a prime number. Therefore, whenever an initial solution is generated by the circle method, a move in the PRS neighborhood is equivalent to a move in the RS neighborhood. Consequently, these neighborhoods always lead to 1-factorizations that are isomorphic to the canonical one.

We may argue if the PTS neighborhood may be used to make the search escape from canonical 1-factorizations. In some cases, this neighborhood suffers from a similar drawback. PTS moves are equivalent to TS moves when there are $2n - 2$ chains of length two in the set $C = \{C_1, \ldots, C_p\}$ computed as explained in Sect. 3.5.5. We could still ask if the neighborhoods PTS and TS are equivalent for some values of $2n$ for which the canonical 1-factorization is perfect. In order to provide an answer to this conjecture, we applied an exhaustive search procedure to check if that was the case for each value of $2n = p+1 \leq 100$, where p is prime. We found that the PTS neighborhood is equivalent to the TS neighborhood for canonical 1-factorizations of K_{2n} for $2n = 4, 6, 12, 14, 20, 30, 38, 54, 60, 62, 68$, and 84.

These results show that, for all the above values of $2n$, any local search heuristic using an initial solution constructed by the circle method will be trapped in a small portion of the search space formed by all 1-factorizations isomorphic to the canonical one. In such cases, applying Vizing's algorithm described in Sect. 2.16 instead of the circle method, or using the GPTS neighborhood, might improve the effectiveness of the local search by exploring a broader portion of the search space.

An experiment using Vizing's algorithm and the circle method to build initial solutions for the traveling tournament problem with predefined venues was performed on instances with 18 and 20 teams. No significant improvements were observed for the instances with 18 teams when Vizing's algorithm was applied because, in this case, neighborhood PTS allows to escape from the canonical 1-factorization. However, for the instances with 20 teams, an improvement of almost 19% on average in the cost function was obtained when Vizing's algorithm replaced the circle method for constructing the initial solution. This experiment revealed that the coloring method based on Vizing's theorem could yield better results than canonical 1-factorizations.

It is open whether neighborhood GPTS is connected. However, the example depicted in Tables 3.1 and 3.2 and Figs. 3.11 and 3.12 shows that the neighborhood GPTS can connect the canonical 1-factorization of K_{12} with a non-isomorphic 1-factorization that is not perfect. Table 3.1 illustrates a perfect 1-factorization of K_{12} for which neighborhoods PTS and TS are equivalent and gives its 11 1-factors.

Figure 3.11 shows the Hamiltonian cycle built using factors F_1 (edges in red) and F_2 (edges in blue) of the perfect 1-factorization of K_{12} in Table 3.1. Figure 3.12 shows the two resulting factors F_1' and F_2' that form two distinct cycles after a move in neighborhood GPTS. The resulting 1-factorization appears in Table 3.2. The edges in bold in some factor F_a' are those that previously belonged to a factor F_b, with $a \neq b$. Thus, neighborhood GPTS can connect the canonical 1-factorization of K_{12} with a non-isomorphic 1-factorization that is not perfect. We recall that the canonical 1-factorization of K_{12} is one of the cases for which both TS and PTS are

Table 3.1 Perfect 1-factorization of K_{12} for which neighborhoods PTS and TS are equivalent

$F_1 = \{(1,12),(2,11),(3,10),(4,9),(5,8),(6,7)\}$
$F_2 = \{(1,10),(2,9),(3,8),(4,7),(5,6),(11,12)\}$
$F_3 = \{(1,8),(2,7),(3,6),(4,5),(9,11),(10,12)\}$
$F_4 = \{(1,6),(2,5),(3,4),(7,11),(8,10),(9,12)\}$
$F_5 = \{(1,4),(2,3),(5,11),(6,10),(7,9),(8,12)\}$
$F_6 = \{(1,2),(3,11),(4,10),(5,9),(6,8),(7,12)\}$
$F_7 = \{(1,11),(2,10),(3,9),(4,8),(5,7),(6,12)\}$
$F_8 = \{(1,9),(2,8),(3,7),(4,6),(5,12),(10,11)\}$
$F_9 = \{(1,7),(2,6),(3,5),(4,12),(8,11),(9,10)\}$
$F_{10} = \{(1,5),(2,4),(3,12),(6,11),(7,10),(8,9)\}$
$F_{11} = \{(1,3),(2,12),(4,11),(5,10),(6,9),(7,8)\}$

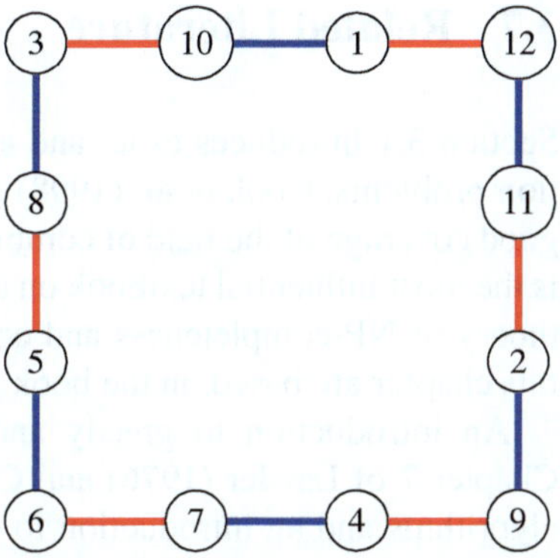

Fig. 3.11 Hamiltonian cycle built using the first (edges in red) and second (edges in blue) factors of the perfect 1-factorization of K_{12} in Table 3.1

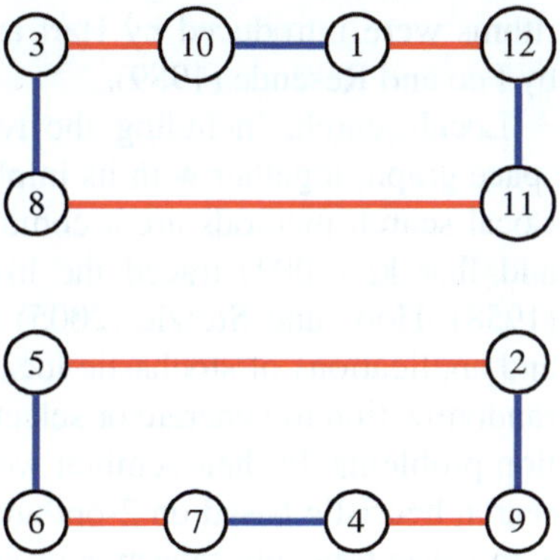

Fig. 3.12 Two cycles formed by the first (edges in red) and second (edges in blue) factors of the 1-factorization of K_{12} in Table 3.2

Table 3.2 1-factorization of K_{12} obtained after the move in neighborhood GPTS. This 1-factorization is no longer perfect (edges in bold in some factor F'_a are those that previously belonged to a factor F_b, with $a \neq b$)

$F'_1 = \{(1, 12), \mathbf{(2, 5)}, (3, 10), (4, 9), (6, 7), \mathbf{(8,11)}\}$
$F'_2 = \{(1, 10), (2, 9), (3, 8), (4, 7), (5, 6), (11, 12)\}$
$F'_3 = \{\mathbf{(1, 6)}, (2, 7), \mathbf{(3, 4)}, \mathbf{(5, 8)}, (9, 11), (10, 12)\}$
$F'_4 = \{\mathbf{(1, 2)}, \mathbf{(3, 6)}, \mathbf{(4, 5)}, (7, 11), (8, 10), (9, 12)\}$
$F'_5 = \{(1, 4), (2, 3), (5, 11), (6, 10), (7, 9), (8, 12)\}$
$F'_6 = \{\mathbf{(1, 8)}, \mathbf{(2, 6)}, (3, 11), (4, 10), (5, 9), (7, 12)\}$
$F'_7 = \{(1, 11), (2, 10), (3, 9), (4, 8), (5, 7), (6, 12)\}$
$F'_8 = \{(1, 9), (2, 8), (3, 7), (4, 6), (5, 12), (10, 11)\}$
$F'_9 = \{(1, 7), \mathbf{(2,11)}, (3, 5), (4, 12), \mathbf{(6, 8)}, (9, 10)\}$
$F'_{10} = \{(1, 5), (2, 4), (3, 12), (6, 11), (7, 10), (8, 9)\}$
$F'_{11} = \{(1, 3), (2, 12), (4, 11), (5, 10), (6, 9), (7, 8)\}$

equivalent, as well as RS and PRS. Although this is not a proof of connectivity, it shows that neighborhood GPTS creates some connections in the search space graph $\mathcal{G} = (F, M)$ associated with the union of neighborhoods HAS, TS, RS, PTS, and PRS.

3.7 Related Literature

Section 3.1 introduces exact and approximate methods to combinatorial optimization problems. Cook et al. (1998) and Papadimitriou and Steiglitz (1982) provided good coverage of the field of combinatorial optimization. Garey and Johnson (1979) is the most influential textbook on computational complexity theory, introducing the theory of NP-completeness and computer intractability. The three first sections of this chapter are based on the book by Resende and Ribeiro (2016b).

An introduction to greedy and semi-greedy algorithms appears in Sect. 3.2. Chapter 7 of Lawler (1976) and Chapter 16 of Cormen et al. (2009) cover greedy algorithms and an introduction to matroid theory. Edmonds (1971) established the connection between weighted matroids and greedy algorithms. Semi-greedy algorithms were introduced by Hart and Shogan (1987) and independently developed by Feo and Resende (1989).

Local search, including the representation of neighborhoods and the search space graph, together with its implementation strategies, is the subject of Sect. 3.3. Local search methods are a common component of most metaheuristics. Yagiura and Ibaraki (2002) traced the history of local search since the work of Croes (1958). Hoos and Stützle (2005) developed a thorough study of the foundations and applications of stochastic local search, methods based on local search that use randomization to generate or select candidate solutions for combinatorial optimization problems. In their seminal work, Lin and Kernighan (1973) developed a local search heuristic based on 2-opt and 3-opt exchanges for approximately solving the symmetric traveling salesman problem, which is one of the best approaches for the problem. Michiels et al. (2007) discussed theoretical aspects of local search.

Chapter 4 of the book by Resende and Ribeiro (2016b) describes enhancements in the implementation of local search methods, such as candidate lists in best-improving strategies and circular search in first-improving strategies; see also Glover and Laguna (1997).

Variable neighborhood descent was described in Mladenovic and Hansen (1997), where the VND acronym was coined; see also Hansen and Mladenović (2003). It was used to implement the local search phase of GRASP heuristics; see, e.g., Martins et al. (2000) for the Steiner problem in graphs and Andreatta and Ribeiro (2002) for the phylogeny problem, among other applications.

The interested reader in the field of metaheuristics and the many existing approaches, some of them summarized in Sect. 3.4, may check the broad and ever-evolving literature on the subject, in particular, the handbooks edited by Reeves (1993b), Glover and Kochenberger (2003), Burke and Kendall (2005, 2014), and Gendreau and Potvin (2010, 2019). Sörensen (2015), de Armas et al. (2022), and Camacho-Villalón et al. (2022, 2023) offer a critical view of the explosion of metaheuristic methods based on metaphors of some natural, bestial, or artificial processes.

The work on optimization by simulated annealing was pioneered by Kirkpatrick et al. (1983), with accounts of later developments and applications in textbooks by van Laarhoven and Aarts (1987) and Aarts and Korst (1989). Convergence and implementation strategies are also discussed in Aarts and Korst (2002) and Henderson et al. (2003). Anagnostopoulos et al. (2006) applied simulated annealing to the traveling tournament problem. Rosati et al. (2022) used a multi-neighborhood simulated annealing approach in the International Timetabling Competition 2021 (Van Bulck et al., 2021), considering the four neighborhoods described in Sects. 3.5.2–3.5.5 and the so-called "partial swap teams phased" neighborhood proposed by the authors.

The seminal papers of Glover (1989, 1990) established the fundamentals, extensions, and uses of tabu search. They provided solid foundations and originated most of the developments from where the field of metaheuristics flourished, as reported in the retrospective by Glover (2022). The reader is referred to the textbook of Glover and Laguna (1997) for a thorough study of tabu search and related ideas. Glover et al. (1993) gave an updated presentation of the fundamental ideas of tabu search with some hints for an efficient implementation. A variety of applications of tabu search with basic modeling suggestions were described by Hertz et al. (1997).

GRASP was proposed by Feo and Resende (1989); see also Feo and Resende (1995). Reviews and tutorials appeared in Ribeiro (2002) and Resende and Ribeiro (2003b, 2005, 2010, 2014). Festa and Resende (2002, 2009) presented reviews of applications of GRASP. The book by Resende and Ribeiro (2016b) consolidated the fundamentals, advances, and extensions of optimization by GRASP.

Path relinking was initially proposed by Glover (1996) as an intensification strategy exploring trajectories connecting elite solutions obtained by tabu search or scatter search. The use of path relinking embedded in a GRASP procedure as an intensification strategy applied to each locally optimal solution was first proposed by Laguna and Martí (1999). Several strategies have been considered and combined with extensions and improvements of path relinking in successful implementations hybridized with tabu search, GRASP, and genetic algorithms; see, e.g., Reeves and Yamada (1998), Canuto et al. (2001), Festa et al. (2002), Ribeiro and Rosseti (2002), Ribeiro et al. (2002), Resende and Ribeiro (2003a), Resende and Ribeiro (2003b), Souza et al. (2003), Martins et al. (2004), Resende and Werneck (2004), Resende and Ribeiro (2005), and Ribeiro and Vianna (2009). Resende and Ribeiro (2005) reviewed these strategies, showing that they involve trade-offs between computation time and solution quality. Chapters 8 and 9 of Resende and Ribeiro (2016b) reviewed path relinking strategies and hybridizations with GRASP.

VNS is a metaheuristic proposed by Mladenovic and Hansen (1997), based on systematic changes in the neighborhood. Other reviews followed it; see Hansen and Mladenović (1999, 2002, 2003).

Iterated local search was proposed in Martin et al. (1991) and Martin and Otto (1996); see Lourenço et al. (2003) for a review. Most of the work using ILS has been done with Markovian random walks, although some studies show that incorporating memory enhances its performance. ILS may also be hybridized with other metaheuristics such as GRASP. For applications of this strategy in sports scheduling, Ribeiro and Urrutia (2007) described an application to the mirrored traveling tournament problem, and Guedes and Ribeiro (2011) to the minimization of the weighted carry-over effects in round-robin tournaments.

Genetic algorithms were first presented in the book of Holland (1975); see also, e.g., Goldberg (1989), Holland (1992), Reeves and Rowe (2002), and Michalewicz (1996), among others. Exemplary implementations require appropriate solution encodings as finite-length data strings and well-tuned strategies for the initial population generation, mate selection, crossover, mutation, and selection; see, e.g., Reeves (1993a, 2003). Genetic algorithms with random keys, or random-key genetic algorithms (RKGA), were first introduced by Bean (1994) for combinatorial optimization problems whose solutions may be represented by permutation vectors. A biased random-key genetic algorithm (BRKGA) differs from an RKGA in the way parents are selected for crossover, see Goncalves and Resende (2011) and Resende and Ribeiro (2016a) for reviews and Buriol et al. (2005), Noronha et al. (2011), Brandão et al. (2015, 2017), and Pinto et al. (2020) for some applications. Path relinking was hybridized with genetic algorithms as a progressive crossover strategy by Ribeiro and Vianna (2009) and with biased random-key genetic algorithms by Andrade et al. (2021) and Ribeiro et al. (2021).

Neighborhoods for round-robin tournaments are the object of Sect. 3.5. Several authors have used neighborhoods home-away swap (HAS), round swap (RS), team swap (TS), partial team swap (PTS), and partial round swap (PRS), although sometimes with different definitions, equivalent or not. Anagnostopoulos et al. (2006) used moves in neighborhoods HAS, RS, TS, PRS, and PTS in their simulated annealing algorithm for the traveling tournament problem. Ribeiro and Urrutia (2007) used moves in neighborhoods HAS, RS, TS, and in a restricted PRS in the local search of their hybrid heuristic for the mirrored traveling tournament problem. They also used ejection chain moves named game rotations as perturbations. Such ejection chain moves are equivalent to moves in the neighborhood PTS, as illustrated in Sect. 3.5. Guedes and Ribeiro (2011) used moves in neighborhoods TS, RS, PRS, and PTS, as well as game rotation, row destruction, and column destruction perturbations in their hybrid ILS with multistart heuristic for the minimum weighted carry-over problem. Costa et al. (2012) used neighborhoods PRS and PTS in the local search of their ILS heuristic for the traveling tournament problem with fixed venues. They also showed that a move in the neighborhood TS (resp. RS) is equivalent to a sequence of moves in the neighborhood PTS (resp. PRS). Di Gaspero and Schaerf (2007) in their composite-neighborhood tabu search approach for the traveling tournament problem used neighborhoods HAS, TS, RS, PTS, and PRS. The last two are seen as repair chains, similar to ejection chains. Rosati et al. (2022) also used neighborhoods TS, RS, PTS, and PRS in their multi-neighborhood simulated annealing for the International Timetabling Competition 2021 (Van Bulck et al., 2021), together with an additional "partial swap teams phased" neighborhood. This neighborhood consists of the application of moves in the PTS neighborhood limited to one of the two SRR schedules of a double round-robin tournament organized in two separate phases (i.e., a phased DRR tournament). Januario and Urrutia (2016) worked with a neighborhood that can be seen as a particular case of GTPS, in which only one chain is allowed to have an arbitrary even length, while the others have length two. Januario et al. (2016a) unified this nomenclature describing all these neighborhoods in terms of edge colorings of a graph.

In Lewis and Thompson (2011), various vertex-coloring algorithms were experimented with for solving round-robin tournament scheduling problems. Local search procedures were explored, and several neighborhood operators were defined for solving real problems with multi-objective optimization techniques.

The connectivity of the tournament space is the subject of Sect. 3.6. Dinitz et al. (1994) computed the complete set of 526,915,620 non-isomorphic 1-factorizations of K_{12}. There are 1,132,835,421,602,062,347 non-isomorphic 1-factorizations of K_{14} reported by Kaski and Ostergard (2009). Kotzig (1963) conjectured that a perfect 1-factorization of K_{2n} exists for every value of $2n > 2$. Wolfe (2009) found a perfect 1-factorization of K_{52} after an exhaustive search that took more than 400 days of computing time. With the discovery by Pike (2019) of a perfect 1-factorization of K_{56}, the smallest graph for which it is unknown whether a perfect 1-factorization exists became K_{64}.

Kobayashi (1989) showed that the values of $2n$ for which the canonical factorization is perfect are precisely the values of $p + 1$ for which $p > 2$ is prime. Januario et al. (2016b) showed that the values of $2n$ for which the canonical 1-factorization is perfect and the PTS and TS neighborhoods coincide are the same for which a riffle permutation permutes all except the first and last of $2n$ distinct integers. Urrutia et al. (2021) coined the term colorful chordless lantern and studied different properties of L-blocking 1-factorizations. They also investigated K-blocking 1-factorizations that do not contain an optimally colored even clique whose cardinality is smaller than the number of 1-factors.

Costa et al. (2012) reported the experiment using Vizing's algorithm to build initial solutions for the traveling tournament problem with predefined venues.

Table 3.3 concludes this chapter by presenting a non-exhaustive collection of applications of metaheuristics to sports scheduling problems. For each metaheuristic, we indicate the problems that motivated the applications and the corresponding references where they appeared. The base metaheuristic sometimes was hybridized with integer programming, constraint programming, or another metaheuristic. Although constraint programming is an efficient approach for scheduling problems in sports with intricate feasibility patterns (Nemhauser and Trick, 1998; Henz, 2001; Benoist et al., 2001; Trick, 2003), entries to its specific applications are omitted from this table; see, e.g., Knust (2022) for an extended list of references about sports scheduling methods and applications.

Table 3.3 Applications of metaheuristics to sports scheduling problems

Metaheuristic	Problem	Reference
Simulated annealing	International Timetabling Competition 2021	Rosati et al. (2022)
	Cricket scheduling (England)	Wright (2018)
	Rugby scheduling (New Zealand)	Johnston and Wright (2014)
	Football scheduling (England)	Kendall et al. (2013)
	MLB umpire scheduling	Trick et al. (2012)
	Baseball scheduling (Korea)	Jeong et al. (2012)
	Traveling tournament problem	Anagnostopoulos et al. (2006)
	Mirrored traveling tournament problem	Van Hentenryck and Vergados (2006)
	Traveling tournament problem	Lim et al. (2006b)
	Scheduling competitions at multiple venues	Lim et al. (2006a)
	Basketball scheduling (New Zealand)	Wright (2006)
	Cricket scheduling (New Zealand)	Wright (2005)
	Cricket scheduling (Australia)	Willis and Terrill (1994)
	Cricket scheduling (England)	Wright (1992)
Tabu search	Indoor football scheduling (Belgium)	Van Bulck et al. (2019)
	Traveling tournament problem	Goerigk and Westphal (2016)
	Volleyball scheduling (Argentina)	Bonomo et al. (2012)
	Traveling tournament problem	Di Gaspero and Schaerf (2007)
	League scheduling	Hamiez and Hao (2001)
	Tennis round-robin scheduling	Della Croce et al. (1999)
	NHL scheduling (Canada and USA)	Costa (1995)
	Cricket scheduling (England)	Wright (1994)
Iterated local search	Traveling umpire problem	Wauters et al. (2014)
	Traveling tournament problem with fixed venues	Costa et al. (2012)
	Weighted carry-over effects	Guedes and Ribeiro (2011)
	Traveling tournament problem with fixed venues	Costa et al. (2008)
	Biobjective referee assignment in amateur leagues	Duarte and Ribeiro (2008)
	Referee assignment in amateur football leagues	Duarte et al. (2007a,b)
	Mirrored traveling tournament problem	Ribeiro and Urrutia (2007)

(continued)

Table 3.3 (continued)

Metaheuristic	Problem	Reference
Local search	Rugby scheduling (Wales)	Lewis and Thompson (2011)
	Hockey scheduling (Finland)	Kyngäs and Nurmi (2009)
	Referee assignment in professional football leagues	Yavuz et al. (2008)
Genetic and memetic algorithms	Football scheduling (England)	While and Kendall (2014)
	Traveling umpire problem	Trick and Yildiz (2012)
	NHL scheduling (Canada and USA)	Craig et al. (2009)
	Scheduling social golfers	Cotta et al. (2006)
	Table tennis scheduling (Germany)	Schönberger et al. (2004), Schönberger and Mattfeld (2000)
Ant colony optimization	Traveling tournament problem	Uthus et al. (2009)
	Maximum value tournament scheduling	Uthus et al. (2008)

References

Aarts E, Korst J (1989) Simulated annealing and Boltzmann machines: a stochastic approach to combinatorial optimization and neural computing. Wiley, Hoboken

Aarts E, Korst J (2002) Selected topics in simulated annealing. In: Ribeiro CC, Hansen P (eds) Essays and surveys in metaheuristics. Kluwer, Dordrecht, pp 1–37

Anagnostopoulos A, Michel L, Van Hentenryck P, Vergados Y (2006) A simulated annealing approach to the traveling tournament problem. J Schedul 9:177–193

Andrade CE, Toso RF, Gonçalves JF, Resende MGC (2021) The multi-parent biased random-key genetic algorithm with implicit path-relinking and its real-world applications. Eur J Oper Res 289:17–30

Andreatta AA, Ribeiro CC (2002) Heuristics for the phylogeny problem. J Heuristics 8:429–447

Bean JC (1994) Genetic algorithms and random keys for sequencing and optimization. ORSA J Comput 6:154–160

Benoist T, Laburthe F, Rottembourg B (2001) Lagrange relaxation and constraint programming collaborative schemes for travelling tournament problems. In: Proceedings of the third international workshop on integration of AI and OR techniques in constraint programming for combinatorial optimization problems, pp 15–26

Bonomo F, Cardemil A, Durán G, Marenco J, Sabán D (2012) An application of the traveling tournament problem: the argentine volleyball league. Interfaces 42:245–259

Brandão JS, Noronha TF, Resende MGC, Ribeiro CC (2015) A biased random-key genetic algorithm for single-round divisible load scheduling. Int Trans Oper Res 22:823–839

Brandão JS, Noronha TF, Resende MGC, Ribeiro CC (2017) A biased random-key genetic algorithm for scheduling heterogeneous multi-round systems. Int Trans Oper Res 27:1061–1077

Buriol LS, Resende MGC, Ribeiro CC, Thorup M (2005) A hybrid genetic algorithm for the weight setting problem in OSPF/IS-IS routing. Networks 46:36–56

Burke E, Kendall G (eds) (2005) Search methodologies: introductory tutorials in optimization and decision support techniques. Springer, Berlin

Burke E, Kendall G (eds) (2014) Search methodologies: introductorial tutorials in optimization and decision support techniques, 2nd edn. Springer, Berlin

Camacho-Villalón CL, Dorigo M, Stützle T (2022) An analysis of why cuckoo search does not bring any novel ideas to optimization. Comput Oper Res 142:105747

Camacho-Villalón CL, Dorigo M, Stützle T (2023) Exposing the grey wolf, moth-flame, whale, firefly, bat, and antlion algorithms: six misleading optimization techniques inspired by bestial metaphors. Int Trans Oper Res 30:2945–2971

Canuto S, Resende MGC, Ribeiro CC (2001) Local search with perturbations for the prize-collecting Steiner tree problem in graphs. Networks 38:50–58

Cook W, Cunningham W, Pulleyblank W, Schrijver A (1998) Combinatorial optimization. Wiley, Hoboken

Cormen T, Leiserson C, Rivest R, Stein C (2009) Introduction to algorithms, 3rd edn. MIT Press, Cambridge

Costa D (1995) An evolutionary tabu search algorithm and the NHL scheduling problem. INFOR Inf Syst Oper Res 33:161–178

Costa FN, Urrutia S, Ribeiro CC (2008) An ILS heuristic for the traveling tournament problem with fixed venues. In: Burke EK, Gendreau M (eds) Proceedings of the 7th international conference on the practice and theory of automated timetabling

Costa FN, Urrutia S, Ribeiro CC (2012) An ILS heuristic for the traveling tournament problem with predefined venues. Ann Oper Res 194:137–150

Cotta C, Dotú I, Fernández AJ, Van Hentenryck P (2006) Scheduling social golfers with memetic evolutionary programming. In: Almeida F, Aguilera MJB, Blum C, Moreno Vega JM, Pérez Pérez M, Roli A, Sampels M (eds) Hybrid metaheuristics. Lecture notes in computer science, vol 4030. Springer, Berlin, pp 150–161

Craig S, While L, Barone L (2009) Scheduling for the national hockey league using a multi-objective evolutionary algorithm. In: Nicholson A, Li X (eds) AI 2009: advances in artificial intelligence. Lecture notes in computer science, vol 5866. Springer, Berlin, pp 381–390

Croes G (1958) A method for solving traveling-salesman problems. Oper Res 6:791–812

de Armas J, Lalla-Ruiz E, Tilahun SL, Voß S (2022) Similarity in metaheuristics: a gentle step towards a comparison. Nat Comput 21:265–287

Della Croce F, Tadei R, Asioli P (1999) Scheduling a round robin tennis tournament under courts and players availability constraints. Ann Oper Res 92:349–361

Di Gaspero L, Schaerf A (2007) A composite-neighborhood tabu search approach to the traveling tournament problem. J Heuristics 13:189–207

Dinitz JH, Garnick DK, McKay BD (1994) There are 526,915,620 nonisomorphic one-factorizations of K_{12}. J Combin Des 2:273–285

Duarte AR, Ribeiro CC (2008) Referee assignment in sports leagues: approximate and exact multi-objective aproaches. In: 19th international conference on multiple criteria decision making, pp 58–60

Duarte AR, Ribeiro CC, Urrutia S (2007a) A hybrid ILS heuristic to the referee assignment problem with an embedded MIP strategy. In: Bartz-Beielstein T, Aguilera MJB, Blum C, Naujoks B, Roli A, Rudolph G, Sampels M (eds) Hybrid metaheuristics. Lecture notes in computer science, vol 4771. Springer, Berlin, pp 82–95

Duarte AR, Ribeiro CC, Urrutia S, Haeusler EH (2007b) Referee assignment in sports leagues. In: Burke EK, Rudová H (eds) Practice and theory of automated timetabling VI. Lecture notes in computer science, vol 3867. Springer, Berlin, pp 158–173

Edmonds J (1971) Matroids and the greedy algorithm. Math Program 1:125–136

Feo TA, Resende MGC (1989) A probabilistic heuristic for a computationally difficult set covering problem. Oper Res Lett 8:67–71

Feo TA, Resende MGC (1995) Greedy randomized adaptive search procedures. J Global Optim 6:109–133

Festa P, Resende MGC (2002) GRASP: an annotated bibliography. In: Ribeiro CC, Hansen P (eds) Essays and surveys in metaheuristics. Kluwer, Dordrecht, pp 325–367

Festa P, Resende MGC (2009) An annotated bibliography of GRASP, part II: applications. Int Trans Oper Res 16:131–172

Festa P, Pardalos P, Resende MGC, Ribeiro CC (2002) Randomized heuristics for the max-cut problem. Optim Methods Softw 7:1033–1058
Garey M, Johnson D (1979) Computers and intractability - a guide to the theory of NP-completeness. W.H. Freeman and Company, New York
Gendreau M, Potvin JY (eds) (2010) Handbook of metaheuristics, 2nd edn. Springer, Berlin
Gendreau M, Potvin JY (eds) (2019) Handbook of metaheuristics, 3rd edn. Springer, Berlin
Glover F (1989) Tabu search - part I. ORSA J Comput 1:190–206
Glover F (1990) Tabu search - part II. ORSA J Comput 2:4–32
Glover F (1996) Tabu search and adaptive memory programming – advances, applications and challenges. In: Barr R, Helgason R, Kennington J (eds) Interfaces in computer science and operations research. Kluwer, Dordrecht, pp 1–75
Glover F (2022) Unforeseen consequences of "Tabu" choices – a retrospective. INFORMS J Comput 34:1306–1308
Glover F, Kochenberger G (eds) (2003) Handbook of metaheuristics. Kluwer, Dordrecht
Glover F, Laguna M (1997) Tabu search. Kluwer, Dordrecht
Glover F, Taillard E, de Werra D (1993) A user's guide to tabu search. In: Glover F, Laguna M, Taillard E, de Werra D (eds) Tabu search, annals of operations research, vol 41, pp 3–28
Goerigk M, Westphal S (2016) A combined local search and integer programming approach to the traveling tournament problem. Ann Oper Res 239:343–354
Goldberg DE (1989) Genetic algorithms in search, optimization and machine learning. Addison-Wesley, Boston
Goncalves JF, Resende MGC (2011) Biased random-key genetic algorithms for combinatorial optimization. J Heuristics 17:487–525
Guedes ACB, Ribeiro CC (2011) A heuristic for minimizing weighted carry-over effects in round robin tournaments. J Schedul 14:655–667
Hamiez JP, Hao JK (2001) Solving the sports league scheduling problem with tabu search. In: Nareyek A (ed) Local search for planning and scheduling. Lecture notes in artificial intelligence, vol 2148, Springer, Berlin, pp 24–36
Hansen P, Mladenović N (1999) An introduction to variable neighbourhood search. In: Voss S, Martello S, Osman I, Roucairol C (eds) Metaheuristics: advances and trends in local search procedures for optimization, Kluwer, Dordrecht, pp 433–458
Hansen P, Mladenović N (2002) Developments of variable neighborhood search. In: Ribeiro CC, Hansen P (eds) Essays and surveys in metaheuristics. Kluwer, Dordrecht, pp 415–439
Hansen P, Mladenović N (2003) Variable neighborhood search. In: Glover F, Kochenberger G (eds) Handbook of metaheuristics. Kluwer, Dordrecht, pp 145–184
Hart J, Shogan A (1987) Semi-greedy heuristics: an empirical study. Oper Res Lett 6:107–114
Henderson D, Jacobson S, Johnson A (2003) The theory and practice of simulated annealing. In: Glover F, Kochenberger G (eds) Handbook of metaheuristics. Kluwer, Dordrecht, pp 287–319
Henz M (2001) Scheduling a major college basketball conference - revisited. Oper Res 49:163–168
Hertz A, Taillard E, de Werra D (1997) Tabu search. In: Aarts E, Lenstra JK (eds) Local search in combinatorial optimization. Wiley, Hoboken, pp 121–143
Holland J (1975) Adaptation in natural and artificial systems. University of Michigan Press, Michigan
Holland J (1992) Genetic algorithms. Sci Am 267:44–50
Hoos H, Stützle T (2005) Stochastic local search: foundations and applications. Elsevier, Amsterdam
Januario T, Urrutia S (2016) A new neighborhood structure for round robin scheduling problems. Comput Oper Res 70:127–139
Januario T, Urrutia S, Ribeiro CC, de Werra D (2016a) Edge coloring: a natural model for sports scheduling. Eur J Oper Res 254:1–8
Januario T, Urrutia S, de Werra D (2016b) Sports scheduling search space connectivity: a riffle shuffle driven approach. Discrete Appl Math 211:113–120
Jeong HM, Kim SW, Choi Y, Kim AJ, Eun J, Kim BJ (2012) Traveling baseball players' problem in Korea. J Korean Phys Soc 61:484–492

Johnston M, Wright M (2014) Prior analysis and scheduling of the 2011 Rugby Union ITM Cup in New Zealand. J Oper Res Soc 65:1292–1300

Kaski P, Ostergard PRJ (2009) There are 1,132,835,421,602,062,347 nonisomorphic one-factorizations of K_{14}. J Combin Des 17:147–159

Kendall G, McCollum B, Cruz FRB, McMullan P, While L (2013) Scheduling English football fixtures: consideration of two conflicting objectives. In: Talbi EG (ed) Hybrid metaheuristics, studies in computational intelligence, vol 434. Springer, Berlin, pp 369–385

Kirkpatrick S, Gelatt C Jr, Vecchi M (1983) Optimization by simulated annealing. Science 220(4598):671–680

Knust S (2022) Classification of literature on sports scheduling. http://www2.inf.uos.de/knust/sportssched/sportlit_class/. Accessed 12 Sept 2022

Kobayashi M (1989) On perfect one-factorization of the complete graph K_{2p}. Graphs Combin 5:351–353

Kotzig A (1963) Hamilton graphs and hamilton circuits. In: Fiedler M (ed) Theory of graphs and its applications: proceedings of the symposium in smolenice, pp 62–82

Kyngäs J, Nurmi K (2009) Scheduling the Finnish major ice hockey league. In: Proceedings of the 2009 IEEE symposium on computational intelligence in scheduling. IEEE, Manhattan, pp 84–89

Laguna M, Martí R (1999) GRASP and path relinking for 2-layer straight line crossing minimization. INFORMS J Comput 11:44–52

Lawler E (1976) Combinatorial optimization: networks and matroids. Holt, Rinehart and Winston, Austin

Lewis R, Thompson J (2011) On the application of graph colouring techniques in round-robin sports scheduling. Comput Oper Res 38:190–204

Lim A, Rodrigues B, Zhang X (2006a) Scheduling sports competitions at multiple venues - revisited. Eur J Oper Res 175:171–186

Lim A, Rodrigues B, Zhang X (2006b) A simulated annealing and hill-climbing algorithm for the traveling tournament problem. Eur J Oper Res 174:1459–1478

Lin S, Kernighan B (1973) An effective heuristic algorithm for the traveling-salesman problem. Oper Res 21:498–516

Lourenço HR, Martin OC, Stutzle T (2003) Iterated local search. In: Glover F, Kochenberger GA (eds) Handbook of metaheuristics. Kluwer, Dordrecht, pp 321–353

Martin OC, Otto SW (1996) Combining simulated annealing with local search heuristics. Ann Oper Res 63:57–75

Martin OC, Otto SW, Felten EW (1991) Large-step markov chains for the traveling salesman problem. Complex Syst 5:299–326

Martins SL, Pardalos PM, Resende MGC, Ribeiro CC (2000) A parallel GRASP for the Steiner tree problem in graphs using a hybrid local search strategy. J Global Optim 17:267–283

Martins SL, Ribeiro CC, Rosseti I (2004) Applications and parallel implementations of metaheuristics in network design and routing. In: Manandhar S, Austin J, Desai U, Oyanagi Y, Talukder AK (eds) Applied computing. Lecture notes in computer science, vol 3285. Springer, Berlin, pp 205–213

Michalewicz Z (1996) Genetic algorithms + data structures = evolution programs. Springer, Berlin

Michiels W, Aarts E, Korst J (2007) Theoretical aspects of local search. Springer, Berlin

Mladenovic N, Hansen P (1997) Variable neighborhood search. Comput Oper Res 34:1097–1100

Nemhauser GL, Trick MA (1998) Scheduling a major college basketball conference. Oper Res 46:1–8

Noronha TF, Resende MGC, Ribeiro CC (2011) A biased random-key genetic algorithm for routing and wavelength assignment. J Global Optim 50:503–518

Papadimitriou C, Steiglitz K (1982) Combinatorial optimization. Prentice-Hall, Hoboken

Pike DA (2019) A perfect one-factorization of K_{56}. arXiv:1810.08734v2. Accessed 10 June 2022

Pinto BQ, Ribeiro CC, Rosseti I, Noronha TF (2020) A biased random-key genetic algorithm for routing and wavelength assignment under a sliding scheduled traffic model. J Global Optim 77:949–973

Reeves C (1993a) Genetic algorithms. In: Reeves C (ed) Modern heuristic techniques for combinatorial problems. Wiley, Hoboken, pp 151–196

Reeves C (ed) (1993b) Modern heuristic techniques for combinatorial problems. Blackwell, Hoboken

Reeves C (2003) Genetic algorithms. In: Glover F, Kochenberger G (eds) Handbook of metaheuristics. Kluwer, Dordrecht, pp 65–82

Reeves C, Rowe J (2002) Genetic algorithms: principles and perspectives. Springer, Berlin

Reeves C, Yamada T (1998) Genetic algorithms, path relinking and the flowshop sequencing problem. Evol Comput 6:45–60

Resende MGC, Ribeiro CC (2003a) A GRASP with path-relinking for private virtual circuit routing. Networks 41:104–114

Resende MGC, Ribeiro CC (2003b) Greedy randomized adaptive search procedures. In: Glover F, Kochenberger G (eds) Handbook of metaheuristics. Kluwer, Dordrecht, pp 219–249

Resende MGC, Ribeiro CC (2005) GRASP with path-relinking: recent advances and applications. In: Ibaraki T, Nonobe K, Yagiura M (eds) Metaheuristics: progress as real problem solvers. Springer, Berlin, pp 29–63

Resende MGC, Ribeiro CC (2010) Greedy randomized adaptive search procedures: advances and applications. In: Gendreau M, Potvin JY (eds) Handbook of metaheuristics, 2nd edn. Springer, Berlin, pp 293–319

Resende MGC, Ribeiro CC (2014) GRASP: greedy randomized adaptive search procedures. In: Burke E, Kendall G (eds) Search methodologies: introductorial tutorials in optimization and decision support systems, 2nd edn. Springer, Berlin, chap 11, pp 287–312

Resende MGC, Ribeiro CC (2016a) Biased-random key genetic algorithms: an advanced tutorial. In: Companion volume - proceedings of the 2016 genetic and evolutionary computation conference - GECCO'16, ACM, New York, pp 483–514

Resende MGC, Ribeiro CC (2016b) Optimization by GRASP: greedy randomized adaptive search procedures. Springer, Berlin

Resende MGC, Werneck RF (2004) A hybrid heuristic for the p-median problem. J Heuristics 10:59–88

Ribeiro CC (2002) GRASP: Une métaheuristique gloutone et probabiliste. In: Teghem J, Pirlot M (eds) Optimisation approchée en recherche opérationnelle, Hermès, pp 153–176

Ribeiro CC, Rosseti I (2002) A parallel GRASP heuristic for the 2-path network design problem. In: Monien B, Feldmann R (eds) Euro-Par 2002 parallel processing. Lecture notes in computer science, vol 2400. Springer, Berlin, pp 922–926

Ribeiro CC, Urrutia S (2007) Heuristics for the mirrored traveling tournament problem. Eur J Oper Res 179:775–787

Ribeiro CC, Vianna DS (2009) A hybrid genetic algorithm for the phylogeny problem using path-relinking as a progressive crossover strategy. Int Trans Oper Res 16:641–657

Ribeiro CC, Uchoa E, Werneck R (2002) A hybrid GRASP with perturbations for the Steiner problem in graphs. INFORMS J Comput 14:228–246

Ribeiro CC, Riveaux JA, Brandão JS (2021) Biased random-key genetic algorithms using path-relinking as a progressive crossover strategy. In: ISMSI'2021: proceedings of the 5th international conference on intelligent systems, metaheuristics & swarm intelligence (ISMSI 2021). ACM, New York, pp 28–36

Rosati RM, Petris M, Di Gaspero L, Schaerf A (2022) Multi-neighborhood simulated annealing for the sports timetabling competition itc2021. J Schedul 25:301–319

Schönberger J, Mattfeld DC (2000) Automated timetable generation for rounds of a table-tennis league. In: Proceedings of the 2000 congress on evolutionary computation, pp 277–284

Schönberger J, Mattfeld DC, Kopfer H (2004) Memetic algorithm timetabling for non-commercial sport leagues. Eur J Oper Res 153:102–116

Sörensen K (2015) Metaheuristics – the metaphor exposed. Int Trans Oper Res 22:1–16

Souza MC, Duhamel C, Ribeiro CC (2003) A GRASP with path-relinking heuristic for the capacitated minimum spanning tree problem. In: Resende MGC, Souza J (eds) Metaheuristics: computer decision making. Kluwer, Dordrecht, pp 627–658

Trick MA (2003) Integer and constraint programming approaches for round robin tournament scheduling. In: Burke E, de Causmaecker P (eds) Practice and theory of automated timetabling IV. Lecture notes in computer science, vol 2740. Springer, Berlin, pp 63–77
Trick MA, Yildiz H (2012) Locally optimized crossover for the traveling umpire problem. Eur J Oper Res 216:286–292
Trick MA, Yildiz H, Yunes T (2012) Scheduling major league baseball umpires and the traveling umpire problem. Interfaces 42:232–244
Urrutia S, de Werra D, Januario T (2021) Recoloring subgraphs of K_{2n} for sports scheduling. Theor Comput Sci 877:36–45
Uthus DC, Riddle PJ, Guesgen HW (2008) Ant colony optimization and the single round robin maximum value problem. In: Dorigo M, Birattari M, Blum C, Clerc M, Stützle T, Winfield AFT (eds) Ant colony optimization and swarm intelligence. Lecture notes in computer science, vol 5217. Springer, Berlin, pp 243–250
Uthus DC, Riddle PJ, Guesgen HW (2009) An ant colony optimization approach to the traveling tournament problem. In: GECCO'09: proceedings of the 11th annual conference on genetic and evolutionary computation. ACM, New York, pp 81–88
Van Bulck D, Goossens DR, Spieksma F (2019) Scheduling a non-professional indoor football league: a tabu search based approach. Ann Oper Res 275:715–730
Van Bulck D, Goossens DR, Beliën J, Davari M (2021) The fifth international timetabling competition (ITC 2021): sports timetabling. In: MathSport international 2021 conference, pp 117–122
Van Hentenryck P, Vergados Y (2006) Traveling tournament scheduling: a systematic evaluation of simulated annealling. In: Integration of AI and OR techniques in constraint programming for combinatorial optimization problems. Lecture notes in computer science, vol 3990. Springer, Berlin, pp 228–243
van Laarhoven P, Aarts E (1987) Simulated annealing: theory and applications. Kluwer, Dordrecht
Wauters T, Malderen SV, Berghe GV (2014) Decomposition and local search based methods for the traveling umpire problem. Eur J Oper Res 238:886–898
While L, Kendall G (2014) Scheduling the English football league with a multi-objective evolutionary algorithm. In: Bartz-Beielstein T, Branke J, Filipič B, Smith J (eds) Parallel problem solving from nature – PPSN XIII. Lecture notes in computer science, vol 8672. Springer, Berlin, pp 842–851
Willis RJ, Terrill BJ (1994) Scheduling the Australian state cricket season using simulated annealing. J Oper Res Soc 45:276–280
Wolfe AJ (2009) A perfect one-factorization of K_{52}. J Combin Des 17:190–196
Wright M (1992) A fair allocation of county cricket opponents. J Oper Res Soc 43:195–201
Wright M (1994) Timetabling county cricket fixtures using a form of tabu search. J Oper Res Soc 45:758–770
Wright M (2005) Scheduling fixtures for New Zealand cricket. IMA J Manag Math 16:99–112
Wright M (2006) Scheduling fixtures for basketball New Zealand. Comput Oper Res 33:1875–1893
Wright M (2018) Scheduling an amateur cricket league over a nine-year period. J Oper Res Soc 69:1854–1862
Yagiura M, Ibaraki T (2002) Local search. In: Pardalos P, Resende M (eds) Handbook of applied optimization, Oxford University Press, Oxford, pp 104–123
Yavuz M, Inan UH, Figlali A (2008) Fair referee assignments for professional football leagues. Comput Oper Res 35:2937–2951

Chapter 4
Integer Programming Approaches

Integer programming has been widely used to solve round-robin tournament scheduling problems, either as standalone formulations or embedded in decomposition strategies. Methods applied to sports scheduling problems include branch-and-bound, branch-and-cut, Benders decomposition, and column generation. They are used to solve scheduling problems in real-life sports leagues and to tackle theoretical problems such as the traveling tournament problem or break minimization.

Hybridizations with other approaches, such as constraint programming and heuristics, are also common in the literature. Some round-robin tournament scheduling problems can be solved by applying a mixed integer programming solver to the complete model. In other cases, decomposition schemes are used.

In this chapter, we illustrate how integer programming can be used to model and solve a typical round-robin tournament scheduling problem. The traveling tournament problem with predefined venues used as the motivation for studying integer programming formulations of round-robin tournament scheduling problems is introduced in Sect. 4.1 and described in Sect. 4.2. Section 4.3 presents some modeling elements (variables and constraints) common to most integer programming formulations of round-robin scheduling problems. Section 4.4 discusses and compares three integer programming formulations for the traveling tournament problem with predefined venues. Finally, Sect. 4.4.4 compares these formulations in terms of the bounds provided by their linear relaxations, and some computational results are given.

4.1 Motivation

In the case of *extramural* round-robin tournaments, each team has its venue at its home city, and each game is played at the venue of either one of the two teams in confrontation. The team that plays at its venue is called the home team and is said

C. C. Ribeiro et al., *Combinatorial Models for Scheduling Sports Tournaments*,
EURO Advanced Tutorials on Operational Research,
https://doi.org/10.1007/978-3-031-37283-4_4

to play a home game, while the other is called the away team and is said to play an away game. A complete schedule determines the round and the venue for each game.

The tournament scheduling problem is often divided into two subproblems. The timetable's construction consists of determining the round in which each game will be played. The home-away pattern (HAP) determines the condition (home or away) in which each team will play in each round. The timetable and the home-away pattern together determine the tournament schedule.

In a *first-break, then-schedule* approach, the subproblem of the second stage consists of building a feasible opponent schedule for a given home-away pattern. A HAP is feasible if there is an opponent schedule compatible with this HAP. The home-away pattern feasibility problem is that of determining whether a given HAP is feasible or not.

In a *first-schedule, then-break* approach, the subproblem of the second stage consists of finding a home-away pattern (with a minimum number of breaks) corresponding to the timetable determined in the first stage. It can be decided in polynomial time whether a timetable for a single round-robin tournament with $2n$ teams can be scheduled with at most $2n$ breaks. A generalization of the break minimization problem where breaks are generalized to encompass any given pair of rounds, not only consecutive rounds (for example, a pair may be formed by the first and last rounds to avoid any team starting and terminating the competition with a home—or away—game), was proved to be NP-complete.

Although most round-robin tournament scheduling problems involve the construction of both the timetable and the home-away pattern, in some cases either the timetable or the HAP may be predefined and known beforehand in some situations.

The problem of scheduling a single round-robin tournament with predefined home-away assignments is the motivation to illustrate integer programming formulations of round-robin tournament scheduling problems. Variants of this problem find interesting applications in real-life leagues whose double round-robin tournaments are divided into two single round-robin phases. Games in the second phase are the same as the first phase, except for the inversion of their venues. Therefore, the games' venues in the second phase are known beforehand and constrained by those in the first phase.

4.2 The Traveling Tournament Problem with Predefined Venues

We assume that the tournament is played by $2n$ teams, indexed by $1, \ldots, 2n$. Each team has its venue in its home city. All teams are initially at their home cities, to which they must return after their last away game. The distance $d_{ij} \geq 0$ from the home city of team i to that of team j is known, for every $i, j = 1, \ldots, 2n$ ($d_{ii} = 0$ for every $i = 1, \ldots, 2n$). A road trip is a sequence of away games played by a team at the venues of its opponents. This team travels from one opponent's venue to the next without returning home. Every sequence of consecutive home (or away) games played by the same team must have at most three games. The venue of each game is known beforehand.

A home-away assignment (HAA) is an assignment of a venue for each game. A HAA is *balanced* if the difference between the number of home games and the number of away games is at most one for every team. If the timetable is predefined (i.e., the round of each game is known beforehand), the HAP and HAA give the same information and determine the schedule.

A set of ordered pairs of teams represents the home-away assignments. The game between teams i and j is represented either by the ordered pair (i, j) or by the ordered pair (j, i). In the first case, the game between teams i and j takes place at the venue of team j; otherwise, at that of team i. For every two teams i and j, either $(i, j) \in$ HAA or $(j, i) \in$ HAA.

The *traveling tournament problem with predefined venues* (TTPPV) consists in finding a compact single round-robin tournament schedule compatible with the set of home-away assignments HAA, such that the total distance traveled by the teams is minimized. No team plays more than three consecutive home games or away games.

A necessary (but not sufficient) feasibility condition for any instance is that

$$\min(h(t), a(t)) \geq \left\lfloor \frac{2n-1}{4} \right\rfloor,$$

where $h(t)$ (resp. $a(t)$) denotes the number of home (resp. away) games of team t, for every team $t = 1, \ldots, 2n$. This condition holds because each team must play at least one home game and one away game in every four consecutive rounds. If the number of consecutive home (and away) games is unbounded, then there is a feasible schedule for every HAA.

Some integer programming preliminaries common to formulations of different round-robin tournament scheduling problems are presented in Sect. 4.3. Three integer programming formulations for the TTPPV are described and discussed in Sects. 4.4.1–4.4.3, with $O(n^3)$, $O(n^4)$, and $O(n^5)$ variables. Their linear relaxations are compared in Sect. 4.4.4. The chapter ends with a discussion of the related literature.

4.3 Preliminaries

Some modeling elements are common to most integer programming formulations of round-robin tournament scheduling problems.

We denote by $2n$ the number of teams and by R the number of rounds in the tournament. Most integer programming formulations make use of the following binary variables:

$$z_{tjk} = \begin{cases} 1, & \text{if team } t \text{ plays at home against team } j \text{ in round } k, \\ 0, & \text{otherwise,} \end{cases}$$

for teams $t, j = 1, \ldots, 2n$ (with $t \neq j$) and rounds $k = 1, \ldots, R$.

The distances traveled can not be directly modeled by variables z_{tjk}. If they are an issue in the problem, then new binary variables have to be used, such as

$$x_{tijk} = \begin{cases} 1, & \text{if team } t \text{ travels from the venue of team } i \text{ to that of } j \text{ in round } k, \\ 0, & \text{otherwise,} \end{cases}$$

for teams $t, i, j = 1, \ldots, 2n$ and rounds $k = 1, \ldots, R$.

As an example, the constraints of a compact double round-robin tournament with $R = 2(2n-1)$ rounds may be formulated using the above variables:

$$\sum_{k=1}^{R=2(2n-1)} z_{tjk} = 1, \qquad t, j = 1, \ldots, 2n : t \neq j, \tag{4.1}$$

$$\sum_{\substack{j=1 \\ j \neq t}}^{2n} (z_{tjk} + z_{jtk}) = 1, \qquad t = 1, \ldots, 2n, \quad k = 1, \ldots, R = 2(2n-1). \tag{4.2}$$

The first set of constraints (4.1) imposes that each team must meet every other team exactly twice, once at home and once away. The second set of constraints (4.2) establishes that every team plays exactly once in each round.

If the length of the round trips and home stands is limited, then the number of distinct round trips and home stands performed by a team is polynomial in the number of teams. This fact usually allows the use of more descriptive variables defining entire round trips or home stands. Variables representing a team's entire sequence of games during the tournament are used in some formulations.

4.4 Integer Programming Formulations

This section discusses three integer programming formulations for the traveling tournament problem with predefined venues (TTPPV). These formulations are compared in terms of the bounds provided by their linear relaxations.

4.4.1 Formulation with $O(n^3)$ Variables

We define the following decision variables:

$$z_{tjk} = \begin{cases} 1, & \text{if team } t \text{ plays at home against team } j \text{ in round } k, \\ 0, & \text{otherwise,} \end{cases}$$

for teams $t, j = 1, \ldots, 2n$ (with $t \neq j$) and rounds $k = 1, \ldots, 2n - 1$, and

$$y_{tij} = \begin{cases} 1, & \text{if team } t \text{ travels from the venue of team } i \text{ to that of team } j, \\ 0, & \text{otherwise,} \end{cases}$$

for teams $t, i, j = 1, \ldots, 2n$.

Since each game occurs exactly once, the travel of each team t between the venues of two teams i and j is performed at most once, in just one of the two directions. Therefore, the y variables are binary. Variables z and y are used in the formulation (4.3)–(4.15) of TTPPV, with $O(n^3)$ variables:

$$\min F_1(z, y) = \sum_{t=1}^{2n} \sum_{i=1}^{2n} \sum_{\substack{j=1 \\ j \neq i}}^{2n} d_{ij} \cdot y_{tij}, \tag{4.3}$$

subject to:

$$\sum_{k=1}^{2n-1} z_{tjk} = 1, \qquad \forall\, (j, t) \in \text{HAA}, \tag{4.4}$$

$$\sum_{k=1}^{2n-1} z_{tjk} = 0, \qquad \forall\, (t, j) \in \text{HAA}, \tag{4.5}$$

$$\sum_{\substack{j=1 \\ j \neq t}}^{2n} (z_{tjk} + z_{jtk}) = 1, \qquad t = 1, \ldots, 2n, \quad k = 1, \ldots, 2n - 1, \tag{4.6}$$

$$y_{tij} \geq z_{it,k-1} + z_{jtk} - 1, \quad t, i, j = 1, \ldots, 2n : t \neq i \neq j, \quad k = 2, \ldots, 2n - 1, \tag{4.7}$$

$$y_{tit} \geq z_{it,k-1} + \sum_{\substack{j=1 \\ j \neq t}}^{2n} z_{tjk} - 1, \qquad t, i = 1, \ldots, 2n : t \neq i, \quad k = 2, \ldots, 2n - 1, \tag{4.8}$$

$$y_{tti} \geq \sum_{\substack{j=1 \\ j \neq t}}^{2n} z_{tj,k-1} + z_{itk} - 1, \qquad t, i = 1, \ldots, 2n : t \neq i, \quad k = 2, \ldots, 2n-1, \tag{4.9}$$

$$y_{tti} \geq z_{it1}, \qquad t, i = 1, \ldots, 2n : t \neq i, \tag{4.10}$$

$$y_{tit} \geq z_{it,2n-1}, \qquad t, i = 1, \ldots, 2n : t \neq i, \tag{4.11}$$

$$\sum_{q=k}^{k+3} \sum_{\substack{j=1 \\ j \neq t}}^{2n} z_{jtq} \leq 3, \qquad t = 1, \ldots, 2n, \quad k = 1, \ldots, 2n-4, \tag{4.12}$$

$$\sum_{q=k}^{k+3} \sum_{\substack{j=1 \\ j \neq t}}^{2n} z_{jtq} \geq 1, \qquad t = 1, \ldots, 2n, \quad k = 1, \ldots, 2n-4, \tag{4.13}$$

$$z_{tjk} \in \{0, 1\}, \qquad t, j = 1, \ldots, 2n : t \neq j, \quad k = 1, \ldots, 2n-1, \tag{4.14}$$

$$y_{tij} \in \{0, 1\}, \qquad t, i, j = 1, \ldots, 2n : i \neq j. \tag{4.15}$$

The objective function (4.3) to be minimized is the total distance traveled by the teams. Constraints (4.4) and (4.5) ensure that each game in HAA occurs precisely once at the appropriate venue. Constraints (4.6) guarantee that each team plays exactly one game in each round. Constraints (4.7) enforce each team t to perform a trip from the home city of team i to that of team j if it plays two consecutive away games against teams i and j, in this order. Constraints (4.8) enforce each team t to perform a trip from the home city of team i to its home city if it has an away game against the latter, followed by a home game in the next round. Constraints (4.9) compel team t to travel from its home city to that of team i to play away against the latter after a home game in the previous round. Constraints (4.10) force team t to travel to the home city of team i if it plays away against the latter in the first round. Constraints (4.11) force team t to return from the home city of team i if it plays away against the latter in the last round. Together, constraints (4.12) and (4.13) imply that any team t can not play more than three home games or less than one home game (and similarly for away games) in any sequence of four consecutive games. Constraints (4.14) and (4.15) define the integrality requirements. The y variables act as generalized upper bounds to constraints (4.7) to (4.11). Since their costs in the objective function to be minimized are non-negative, they will always assume the minimal possible value, either 0 or 1. Therefore, the integrality requirements on the y variables may be substituted by lower and upper bounds, replacing constraints (4.15).

This formulation has $O(n^2)$ constraints of types (4.4), (4.5), (4.6), (4.10), (4.11), (4.12), and (4.13), $O(n^3)$ of types (4.8) and (4.9), and $O(n^4)$ of type (4.7).

4.4.2 Formulation with $O(n^4)$ Variables

We now reformulate the problem as a network flow model in which each team corresponds to a different commodity. We recall that $h(t)$ denotes the number of home games played by team $t = 1, \ldots, 2n$. We also introduce a dummy round $2n$ to express that every team must return to its home city after its last away game, where it stays until the end of the tournament.

The new decision variables are:

$$x_{tijk} = \begin{cases} 1, & \text{if team } t \text{ travels from the venue of team } i \text{ to that of } j \text{ in round } k, \\ 0, & \text{otherwise,} \end{cases}$$

for teams $t, i, j = 1, \ldots, 2n$ and rounds $k = 1, \ldots, 2n$.

If a team t plays two consecutive home games in rounds k and $k+1$, then $x_{ttt,k+1} = 1$ as if there was a fictitious zero-cost trip leaving from and arriving at the home city of team t to play in round $k+1$. The new x variables are used to build formulation (4.16)–(4.25) with $O(n^4)$ variables:

$$\min F_2(x) = \sum_{t=1}^{2n} \sum_{i=1}^{2n} \sum_{j=1}^{2n} \sum_{k=1}^{2n} d_{ij} \cdot x_{tijk}, \tag{4.16}$$

subject to:

$$\sum_{k=1}^{2n-1} \sum_{i=1}^{2n} x_{tijk} = 1, \qquad \forall (t, j) \in \text{HAA}, \tag{4.17}$$

$$\sum_{j=1}^{2n} x_{ttj1} = 1, \qquad t = 1, \ldots, 2n, \tag{4.18}$$

$$\sum_{\substack{i=1 \\ i \neq t}}^{2n} \sum_{j=1}^{2n} x_{tij1} = 0, \qquad t = 1, \ldots, 2n, \tag{4.19}$$

$$\sum_{j=1}^{2n} x_{tijk} = \sum_{j=1}^{2n} x_{tji,k-1}, \qquad t, i = 1, \ldots, 2n, \quad k = 2, \ldots, 2n, \tag{4.20}$$

$$\sum_{\substack{j=1\\j\neq t}}^{2n}\sum_{i=1}^{2n} x_{jitk} = \sum_{i=1}^{2n} x_{titk}, \qquad t = 1,\ldots,2n, \quad k = 1,\ldots,2n-1, \tag{4.21}$$

$$\sum_{i=1}^{2n}\sum_{k=1}^{2n} x_{titk} = h(t) + 1, \qquad t = 1,\ldots,2n, \tag{4.22}$$

$$\sum_{q=k}^{k+3}\sum_{\substack{j=1\\j\neq t}}^{2n}\sum_{i=1}^{2n} x_{tijq} \leq 3, \qquad t = 1,\ldots,2n, \quad k = 1,\ldots,2n-4, \tag{4.23}$$

$$\sum_{q=k}^{k+3}\sum_{\substack{j=1\\j\neq t}}^{2n}\sum_{i=1}^{2n} x_{tijq} \geq 1, \qquad t = 1,\ldots,2n, \quad k = 1,\ldots,2n-4, \tag{4.24}$$

$$x_{tijk} \in \{0,1\}, \qquad t,i,j,k = 1,\ldots,2n. \tag{4.25}$$

The objective function (4.16) to be minimized is the total distance traveled by the teams. Constraints (4.17) ensure that each game in HAA is scheduled precisely once by forcing the away team to travel to meet the home team. Constraints (4.18) and (4.19) force every team t to perform a trip starting from its home city in the first round. If team t plays at home in the first round, this will be a fictitious trip ending at its home city. Constraints (4.19) are redundant and unnecessary in the integer programming formulation, but they contribute to strengthening the lower bounds provided by the linear relaxation. Constraints (4.20) require that if team t leaves facility i in round k, it must have traveled to facility i in the previous round. Constraints (4.21) state that team t is at home in round k if and only if another team travels to play at its home city. Constraints (4.22) compel team t to return to its home city at the end of the tournament, represented by the dummy round indexed by $2n$. Since team t plays $h(t)$ home games during the tournament, these constraints enforce it to be at home at the end of the tournament by imposing that it has to perform $h(t)+1$ trips to its home city. Constraints (4.23) establish that team t can not play more than three consecutive away games, while constraints (4.24) guarantee that team t can not play more than three consecutive home games. Constraints (4.25) are the integrality requirements.

This formulation has $O(n)$ constraints of types (4.18), (4.19), and (4.22), $O(n^2)$ of types (4.17), (4.21), (4.23), and (4.24), and $O(n^3)$ of type (4.20).

4.4.3 Formulation with $O(n^5)$ Variables

This formulation considers variables associated with complete road trips. The variables represent road trips of different sizes, giving a more direct representation of the problem. We recall that the problem's constraints require that no team should play more than three consecutive home games or three consecutive away games, which limits the number of variables in this formulation. Three new types of more descriptive decision variables are defined and used in this third formulation:

$$w^1_{tik} = \begin{cases} 1, & \text{if team } t \text{ starts, in round } k, \text{ a road trip visiting team } i \text{ and} \\ & \text{returning home in round } k+1 \text{ (with } t \neq i\text{)}, \\ 0, & \text{otherwise;} \end{cases}$$

$$w^2_{tijk} = \begin{cases} 1, & \text{if team } t \text{ starts, in round } k, \text{ a road trip visiting first team } i, \text{ then} \\ & \text{team } j, \text{ and returning home in round } k+2 \text{ (with } t \neq i \neq j\text{)}, \\ 0, & \text{otherwise;} \end{cases}$$

$$w^3_{tij\ell k} = \begin{cases} 1, & \text{if team } t \text{ starts, in round } k, \text{ a road trip visiting first team } i, \\ & \text{then team } j, \text{ next team } \ell, \text{ and returning home in round} \\ & k+3 \text{ (with } t \neq i \neq j \neq \ell\text{)}, \\ 0, & \text{otherwise.} \end{cases}$$

Two dummy rounds (indexed by -1 and 0) are created to simplify the formulation. The variables corresponding to every road trip starting in any of these dummy rounds are set to 0. The additional costs c_{ij}, c_{ijm}, and c_{ijml} represent the costs of road trips of length one, two, and three performed by team i, respectively:

$$c_{ij} = d_{ij} + d_{ji}, \tag{4.26}$$

$$c_{ijm} = d_{ij} + d_{jm} + d_{mi}, \tag{4.27}$$

$$c_{ijm\ell} = d_{ij} + d_{jm} + d_{m\ell} + d_{\ell i}. \tag{4.28}$$

The new variables are used to reformulate TTPPV as the model (4.29)–(4.34) that follows below, with $O(n^5)$ variables:

$$\min F_3(w^1, w^2, w^3) = \sum_{k=1}^{2n-1} \sum_{i=1}^{2n} \sum_{\substack{j=1 \\ (i,j)\in \text{HAA}}}^{2n} [c_{ij} \cdot w^1_{ijk} +$$

$$\sum_{\substack{m=1 \\ (i,m)\in \text{HAA} \\ m\neq j}}^{2n} (c_{ijm} \cdot w^2_{ijmk} + \sum_{\substack{\ell=1 \\ (i,\ell)\in \text{HAA} \\ \ell\neq j\neq m}}^{2n} c_{ijm\ell} \cdot w^3_{ijm\ell k})], \tag{4.29}$$

subject to:

$$\sum_{i=1}^{2n} \sum_{\substack{j=1 \\ (i,j)\in \mathrm{HAA}}}^{2n} [\sum_{k\in\{-1,0\}} w_{ijk}^1 + \sum_{\substack{m=1 \\ (i,m)\in \mathrm{HAA} \\ m\neq j}}^{2n} (\sum_{k\in\{-1,0,2n-1\}} w_{ijmk}^2 + \sum_{\substack{\ell=1 \\ (i,\ell)\in \mathrm{HAA} \\ \ell\neq j\neq m}}^{2n} \sum_{k\in\{-1,0,2n-2,2n-1\}} w_{ijm\ell k}^3)] = 0, \tag{4.30}$$

$$\sum_{k=1}^{2n-1} \{w_{ijk}^1 + \sum_{\substack{m=1 \\ (i,m)\in \mathrm{HAA} \\ m\neq j}}^{2n} [(w_{ijmk}^2 + w_{imjk}^2) + \sum_{\substack{\ell=1 \\ (i,\ell)\in \mathrm{HAA} \\ \ell\neq j\neq m}}^{2n} (w_{ijm\ell k}^3 + w_{imj\ell k}^3 + w_{im\ell jk}^3)]\} = 1, \qquad \forall (i,j) \in \mathrm{HAA}, \tag{4.31}$$

$$\sum_{\substack{j=1 \\ (i,j)\in \mathrm{HAA}}}^{2n} \{w_{ijk}^1 + \sum_{\substack{m=1 \\ (i,m)\in \mathrm{HAA} \\ m\neq j}}^{2n} [(w_{ijmk}^2 + w_{ijm,k-1}^2) + \sum_{\substack{\ell=1 \\ (i,\ell)\in \mathrm{HAA} \\ \ell\neq j\neq m}}^{2n} (w_{ijm\ell k}^3 + w_{ijm\ell,k-1}^3 + w_{ijm\ell,k-2}^3)]\} + \sum_{\substack{j=1 \\ (j,i)\in \mathrm{HAA}}}^{2n} \{w_{jik}^1 + \sum_{\substack{m=1 \\ (j,m)\in \mathrm{HAA} \\ m\neq i}}^{2n} [(w_{jimk}^2 + w_{jmi,k-1}^2) + \sum_{\substack{\ell=1 \\ (j,\ell)\in \mathrm{HAA} \\ \ell\neq i\neq m}}^{2n} (w_{jim\ell k}^3 + w_{jmi\ell,k-1}^3 + w_{jm\ell i,k-2}^3)]\} = 1, \qquad i = 1, \ldots, 2n, \quad k = 1, \ldots, 2n-1, \tag{4.32}$$

$$\sum_{\substack{j=1 \\ (i,j)\in \mathrm{HAA}}}^{2n} \{w_{ijk}^1 + w_{ij,k+1}^1 + \sum_{\substack{m=1 \\ (i,m)\in \mathrm{HAA} \\ m\neq j}}^{2n} [w_{ijm,k-1}^2 + w_{ijmk}^2 + w_{ijm,k+1}^2 + \sum_{\substack{\ell=1 \\ (i,\ell)\in \mathrm{HAA} \\ \ell\neq j\neq m}}^{2n} (w_{ijm\ell,k-2}^3 + w_{ijm\ell,k-1}^3 + w_{ijm\ell k}^3 + w_{ijm\ell,k+1}^3)]\} \leq 1,$$

$$i = 1, \ldots, 2n, \quad k = 1, \ldots, 2n-2, \tag{4.33}$$

$$\sum_{q=k}^{k+3} \sum_{\substack{j=1 \\ (i,j)\in \text{HAA}}}^{2n} \{w^1_{ijq} + \sum_{\substack{m=1 \\ (i,m)\in \text{HAA} \\ m\neq j}}^{2n} [w^2_{ijmq} + w^2_{ijm,q-1} +$$

$$\sum_{\substack{\ell=1 \\ (i,\ell)\in \text{HAA} \\ \ell\neq j\neq m}}^{2n} (w^3_{ijm\ell q} + w^3_{ijm\ell,q-1} + w^3_{ijm\ell,q-2})]\} \geq 1,$$

$$i = 1, \ldots, 2n, \quad k = 1, \ldots, 2n-4, \tag{4.34}$$

$$w^1_{ijk} \in \{0, 1\}, \qquad i, j = 1, \ldots, 2n : i \neq j, \quad k = -1, \ldots, 2n-1, \tag{4.35}$$

$$w^2_{ijmk} \in \{0, 1\}, \qquad i, j, m = 1, \ldots, 2n : i \neq j \neq m, \quad k = -1, \ldots, 2n-1, \tag{4.36}$$

$$w^3_{ijm\ell k} \in \{0, 1\}, \quad i, j, m, \ell = 1, \ldots, 2n : i \neq j \neq m \neq \ell, \quad k = -1, \ldots, 2n-1. \tag{4.37}$$

Once again, the total distance traveled is the objective function (4.29) to be minimized. Constraints (4.30) set to zero all variables associated with road trips starting at the dummy rounds -1 and 0, road trips of size two and three starting in the last round, and round trips of size three starting in round $2n-2$. Constraints (4.31) ensure that each game occurs precisely once. They express that each game $(i, j) \in$ HAA should be played in a road trip of team i formed by one, two, or three away games. Constraints (4.32) enforce that team i is either playing an away game or another team is visiting it in each round. The sum of all variables associated with road trips of team i (which include round k) and road trips of other teams that visit team i in round k is set to one. Constraints (4.33) forbid team i to be engaged in simultaneous or consecutive (i.e., without returning to its home city) road trips in round k. Constraints (4.34) state that team i must be outside its home city to play away at least once every four consecutive rounds. Constraints (4.35)–(4.37) guarantee the integrality requirements.

Although the number of variables increases compared to the previous formulation, the number of constraints is quite smaller: this formulation has only $O(n^2)$ constraints of types (4.31)–(4.34).

4.4.4 Linear Relaxation Bounds and Computational Results

We refer to the formulations with $O(n^3)$, $O(n^4)$ and $O(n^5)$ variables, respectively, by models N3, N4 and N5. We denote by $\overline{\text{N3}}$, $\overline{\text{N4}}$ and $\overline{\text{N5}}$, respectively, their linear relaxations. Let $LB3$, $LB4$, and $LB5$ be the linear relaxation bounds provided by the respective formulations N3, N4, and N5.

We state two theorems that provide comparisons between the formulations concerning their linear relaxation bounds $LB3$, $LB4$, and $LB5$. The first theorem compares formulations N3 and N4, while the second compares formulations N4 and N5.

Theorem 4.1 $LB4 \geq LB3$ *holds for every instance of the TTPPV.* ■

Theorem 4.2 $LB5 \geq LB4$ *holds for every instance of the TTPPV.* ■

In order to compare the quality of the proposed integer programming formulations, 20 home-away assignments were created for each national league (NL) and circular (CIRC) instance of the set of benchmark instances of the traveling tournament problem. Ten instances in each group have their home-away assignments randomly determined. The other ten instances correspond to balanced home-away assignments: the venues of all games were randomly selected, and an iterative method was used to balance the home-away assignments.

The linear relaxations of the three formulations described in Sects. 4.4.1–4.4.3 were solved for all instances. The results are summarized in Tables 4.1 and 4.2 for the NL instances and Tables 4.3 and 4.4 for the CIRC instances. For each value of $2n$, the second column shows the number of feasible instances (since not every HAA corresponds to a feasible instance). The three following columns of these tables give the average lower bounds provided by models N3, N4, and N5 for the feasible instances. The four last columns give the average and maximum gaps between the bounds $LB3$ and $LB5$, and between the bounds $LB4$ and $LB5$. The relative gaps are given by $100 \cdot (LB5 - LB3)/LB3$ and $100 \cdot (LB5 - LB4)/LB4$, i.e., by how much $LB5$ improves upon $LB3$ and upon $LB4$, respectively.

The average bounds $LB5$ provided by formulation N5 are far better than those provided by N3. Considering, e.g., the random CIRC instances with $2n = 20$, the average bound $LB5$ is approximately 41.7 times greater than the average bound $LB3$. The average bound $LB5$ also improves upon those given by model N4.

Table 4.1 Linear relaxation lower bounds for the random NL instances

					$(LB5 - LB3)/LB3)$		$(LB5 - LB4)/LB4)$	
Teams $2n$	Feasible instances	$LB3$	$LB4$	$LB5$	Avg. (%)	Max. (%)	Avg. (%)	Max. (%)
4	10	2064.92	2090.67	5227.70	153.16	191.59	150.05	188.57
6	7	3716.91	5503.67	13756.31	273.03	312.10	154.32	210.15
8	10	4558.85	7600.34	21675.06	401.57	550.88	214.08	350.24
10	6	3858.40	7175.11	32390.32	740.40	788.92	357.24	430.72
12	9	5989.87	14520.32	57930.83	877.94	984.17	333.60	514.68
14	7	6658.78	15968.46	103499.29	1455.36	1533.61	549.76	596.60
16	9	7280.00	24599.24	135829.89	1805.62	1996.64	484.61	607.84

Table 4.2 Linear relaxation lower bounds for the balanced NL instances

Teams	Feasible				$(LB5 - LB3)/LB3)$		$(LB5 - LB4)/LB4)$	
					Avg.	Max.	Avg.	Max.
$2n$	instances	$LB3$	$LB4$	$LB5$	(%)	(%)	(%)	(%)
4	10	2063.90	2090.67	5097.50	147.00	191.24	143.82	186.80
6	10	2992.80	3896.80	14134.90	372.42	396.69	262.73	274.92
8	10	3456.92	4991.71	21278.91	515.59	536.58	326.28	340.37
10	10	3704.52	6218.44	32573.49	779.29	809.42	423.82	440.13
12	10	5377.26	9482.18	58432.26	986.73	1017.92	516.23	533.14
14	10	6533.32	15325.50	103722.30	1487.59	1529.30	576.80	593.96
16	10	6440.18	19114.40	135563.90	2004.97	2040.95	609.22	621.46

Table 4.3 Linear relaxation lower bounds for the random CIRC instances

Teams	Feasible				$(LB5 - LB3)/LB3)$		$(LB5 - LB4)/LB4)$	
					Avg.	Max.	Avg.	Max.
$2n$	instances	$LB3$	$LB4$	$LB5$	(%)	(%)	(%)	(%)
4	10	4.94	5.33	12.60	154.50	180.00	136.25	162.50
6	7	8.50	13.6	35.90	329.52	410.81	167.09	224.07
8	8	14.53	29.92	73.78	431.45	702.81	156.29	282.81
10	7	12.81	33.43	132.76	971.62	1197.72	303.46	370.40
12	10	15.03	49.54	214.23	1402.89	1667.04	358.40	450.00
14	8	16.35	60.65	326.62	1907.82	2179.34	439.15	455.73
16	10	18.90	83.77	471.70	2488.63	2860.97	502.15	596.29
18	8	19.24	91.67	649.38	3294.49	3528.10	609.96	655.17
20	8	20.83	113.73	869.29	4102.11	4252.44	684.96	729.35

Table 4.4 Linear relaxation lower bounds for the balanced CIRC instances

Teams	Feasible				$(LB5 - LB3)/LB3)$		$(LB5 - LB4)/LB4)$	
					Avg.	Max.	Avg.	Max.
$2n$	instances	$LB3$	$LB4$	$LB5$	(%)	(%)	(%)	(%)
4	10	4.94	5.33	13.40	171.00	180.00	151.25	162.50
6	10	6.92	10.80	34.70	402.08	444.19	221.30	233.33
8	10	8.74	18.29	71.70	720.81	747.78	292.11	297.40
10	10	10.49	27.78	131.07	1150.98	1194.31	371.84	382.40
12	10	12.13	39.27	213.57	1659.98	1684.34	443.80	451.70
14	10	14.15	52.77	326.30	2206.79	2250.43	518.35	527.89
16	10	16.07	68.27	466.00	2800.43	2850.29	582.62	595.80
18	10	18.02	85.76	647.30	3490.46	3548.08	654.74	666.44
20	10	20.02	105.26	867.05	4230.67	4292.39	723.70	735.05

Considering the same random CIRC instances with $2n = 20$, the average bound $LB5$ is approximately 7.6 times greater than the average bound $LB4$.

4.5 Related Literature

Integer programming is a useful tool to model and solve round-robin tournament scheduling problems; see, e.g., the textbook of Wolsey (2021). Integer programming was used for solving scheduling problems in football leagues; see, e.g., Bartsch et al. (2006), Della Croce and Oliveri (2006), Noronha et al. (2007), Ribeiro and Urrutia (2007, 2012), Durán et al. (2007), Rasmussen (2008), Goossens and Spieksma (2009), Durán et al. (2009), Durán et al. (2012), Alarcón et al. (2014), Durán et al. (2017), Goossens (2018), and Durán et al. (2022). Integer programming was also applied in the solution of the traveling tournament problem; see, e.g., Easton et al. (2003), Rasmussen and Trick (2006), Melo et al. (2009), and Irnich (2010). Urrutia et al. (2007) used integer programming to derive improved lower bounds to the traveling tournament problem. It was also used to solve many other tournament scheduling applications; see, e.g., Nemhauser and Trick (1998), Trick (2001), Urban and Russell (2003), and Farmer et al. (2007). Case studies and applications to other sports scheduling problems are reported, e.g., in Kendall et al. (2010), Ribeiro (2012), and Knust (2022), see also Sect. 5.4.

Section 4.1 motivates the use of decomposition approaches, which have been introduced first in Sect. 2.7. First-break, then-schedule and first-schedule, then-break approaches have been reviewed by Kendall et al. (2010) in Section 3.1 of their annotated bibliography. Trick (2001) pioneered the use of first-schedule, then-break approaches in sports timetabling, and coined this terminology. Algorithms based on a first-break, then-schedule approach can be found in Nemhauser and Trick (1998), Henz (2001), Rasmussen and Trick (2006, 2007), Rasmussen (2008), Knust and Lücking (2009), and Ribeiro and Urrutia (2012), among others.

For complexity results on break minimization, see Goossens and Spieksma (2011). The break minimization problem (Régin, 2001) deals with minimizing the number of breaks in the schedule. In contrast, timetable-constrained distance minimization problems (Rasmussen and Trick, 2006) call for minimizing the total distance traveled.

A necessary condition for the feasibility of a home-away pattern is given in Miyashiro et al. (2003). Constructing a timetable compatible with a given home-away pattern and optimizing a particular objective appears as a subproblem in several approaches to solve real-life scheduling problems; see, e.g., Nemhauser and Trick (1998).

Section 4.2 introduces the traveling tournament problem with predefined venues. It has applications in real-life leagues whose double round-robin tournaments are divided into two single round-robin phases. Games in the second phase are scheduled in the same rounds as in the first, with interchanged venues. Therefore, the venues of the second phase games are known beforehand and determined by those of the first phase. The Chilean professional football league (Noronha et al., 2007) and the German table tennis federation of Lower Saxony (Knust, 2007) hold tournaments with this design. Home-away assignments (HAA) were defined in Knust and von Thaden (2006).

The work described in Sects. 4.3 and 4.4 appeared originally in Melo et al. (2009), where the proofs of Theorems 1 and 2 were given. The comparison of the quality of the integer programming formulations was based on 20 home-away assignments created for each benchmark national league (NL) and circular (CIRC) instance of the traveling tournament problem in Trick (2020); see also Van Bulck et al. (2020) for a taxonomy and unified data format for round-robin sports timetabling. Ten instances in each group have their home-away assignments randomly determined. The other ten instances correspond to balanced home-away assignments: the venues of all games were selected randomly, and the iterative method in Knust and von Thaden (2006) was used to balance the home-away assignments.

A similar comparative study was developed by Van Doornmalen et al. (2022), where the authors considered the problem of finding an optimal schedule for compact single round-robin tournaments and three formulations. The first is a traditional formulation using $O(n^3)$ variables and constraints, also studied by Trick (2003) and Briskorn and Drexl (2009). The other two formulations are based on different encodings of a schedule: a matching and a permutation formulation. The strengths of the linear programming relaxations of the three models were compared.

References

Alarcón F, Durán G, Guajardo M (2014) Referee assignment in the Chilean football league using integer programming and patterns. Int Trans Oper Res 21:415–438

Bartsch T, Drexl A, Kröger S (2006) Scheduling the professional soccer leagues of Austria and Germany. Comput Oper Res 33:1907–1937

Briskorn D, Drexl A (2009) IP models for round robin tournaments. Comput Oper Res 36:837–852

Della Croce F, Oliveri D (2006) Scheduling the Italian football league: an ILP-based approach. Comput Oper Res 33:1963–1974

Durán G, Guajardo M, Miranda J, Sauré D, Souyris S, Weintraub A, Wolf R (2007) Scheduling the Chilean soccer league by integer programming. Interfaces 37:539–552

Durán G, Guajardo M, Weintraub A, Wolf R (2009) O.R. & Soccer: scheduling the Chilean league using mathematical programming. OR/MS Today 36:42–47

Durán G, Guajardo M, Wolf-Yadlin R (2012) Operations research techniques for scheduling Chile's second division soccer league. Interfaces 42:273–285

Durán G, Guajardo M, Sauré D (2017) Scheduling the South American qualifiers to the 2018 FIFA World Cup by integer programming. Eur J Oper Res 262:1109–1115

Durán G, Guajardo M, Zamorano G (2022) Mathematical models for rescheduling Ecuador's 2020 professional football league season disrupted by COVID-19. Soc Econ 44:360–377

Easton K, Nemhauser GL, Trick MA (2003) Solving the travelling tournament problem: a combined integer programming and constraint programming approach. In: Burke EK, de Causmaecker P (eds) Practice and theory of automated timetabling IV. Lecture notes in computer science, vol 2740. Springer, Berlin, pp 100–109

Farmer A, Smith JS, Miller LT (2007) Scheduling umpire crews for professional tennis tournaments. Interfaces 37:187–196

Goossens DR (2018) Optimization in sports league scheduling: experiences from the Belgian Pro League soccer. Commun Comput Inf Sci 884:3–19

Goossens DR, Spieksma FCR (2009) Scheduling the Belgian soccer league. Interfaces 39:109–118

Goossens DR, Spieksma FCR (2011) Breaks, cuts, and patterns. Oper Res Lett 39:428–432

Henz M (2001) Scheduling a major college basketball conference - revisited. Oper Res 49:163–168

Irnich S (2010) A new branch-and-price algorithm for the traveling tournament problem. Eur J Oper Res 204:218–228

Kendall G, Knust S, Ribeiro CC, Urrutia S (2010) Scheduling in sports: an annotated bibliography. Comput Oper Res 37:1–19

Knust S (2007) Scheduling non-professional table-tennis leagues. Technical Report 270, Osnabrücker Schriften zur Mathematik

Knust S (2022) Classification of literature on sports scheduling. Online reference at http://www2.inf.uos.de/knust/sportssched/sportlit_class/, last access on September 12, 2022

Knust S, Lücking D (2009) Minimizing costs in round robin tournaments with place constraints. Comput Oper Res 36:2937–2943

Knust S, von Thaden M (2006) Balanced home-away assignments. Discrete Optim 3:354–365

Melo RA, Urrutia S, Ribeiro CC (2009) The traveling tournament problem with predefined venues. J Sched 12:607–622

Miyashiro R, Iwasaki H, Matsui T (2003) Characterizing feasible pattern sets with a minimum number of breaks. In: Goos G, Hartmanis J, van Leeuwen J (eds) Practice and theory of automated timetabling IV. Lecture notes in computer science, vol 2740. Springer, Berlin, pp 78–99

Nemhauser GL, Trick MA (1998) Scheduling a major college basketball conference. Oper Res 46:1–8

Noronha TF, Ribeiro CC, Durán G, Souyris S, Weintraub A (2007) A branch-and-cut algorithm for scheduling the highly-constrained Chilean soccer tournament. In: Burke EK, Rudová H (eds) Practice and theory of automated timetabling VI. Lecture notes in computer science, vol 3867. Springer, Berlin, pp 174–186

Rasmussen RV (2008) Scheduling a triple round robin tournament for the best Danish soccer league. Eur J Oper Res 185:795–810

Rasmussen RV, Trick MA (2006) The timetable constrained distance minimization problem. In: Beck JC, Smith BM (eds) Integration of AI and OR techniques in constraint programming for combinatorial optimization problems. Lecture notes in computer science, vol 3990. Springer, Berlin, pp 167–181

Rasmussen RV, Trick MA (2007) A Benders approach for the constrained minimum break problem. Eur J Oper Res 177:198–213

Régin J (2001) Minimization of the number of breaks in sports scheduling problems using constraint programming. DIMACS Ser Discrete Math Theor Comput Sci 57:115–130

Ribeiro CC (2012) Sports scheduling: problems and applications. Int Trans Oper Res 19:201–226

Ribeiro CC, Urrutia S (2007) Scheduling the Brazilian soccer tournament with fairness and broadcast objectives. In: Burke EK, Rudová H (eds) Practice and theory of automated timetabling VI. Lecture notes in computer science, vol 3867. Springer, Berlin, pp 147–157

Ribeiro CC, Urrutia S (2012) Scheduling the Brazilian soccer tournament: solution approach and practice. Interfaces 42:260–272

Trick MA (2001) A schedule-then-break approach to sports timetabling. In: Burke EK, Erben E (eds) The 3rd international conference on the practice and theory of automated timetabling. Lecture notes in computer science, vol 2079. Springer, Berlin, pp 242–252

Trick MA (2003) Integer and constraint programming approaches for round robin tournament scheduling. In: Burke E, de Causmaecker P (eds) Practice and theory of automated timetabling IV. Lecture notes in computer science, vol 2740. Springer, Berlin, pp 63–77

Trick MA (2020) Challenge traveling tournament instances. Online reference at http://mat.gsia.cmu.edu/TOURN/, last visited on June 5, 2022

Urban T, Russell R (2003) Scheduling sports competitions on multiple venues. Eur J Oper Res 148:302–311

Urrutia S, Ribeiro CC, Melo RA (2007) A new lower bound to the traveling tournament problem. In: Proceedings of the 2007 IEEE symposium on computational intelligence in scheduling. IEEE, pp 15–18

Van Bulck D, Goossens DR, Schönberger J, Guajardo M (2020) RobinX: a three-field classification and unified data format for round-robin sports timetabling. Eur J Oper Res 280:568–580

Van Doornmalen J, Hojny C, Lambers R, Spieksma FCR (2022) Integer programming formulations for compact single round robin tournaments. In: Causmaecker PD, Özcan E, Berghe GV (eds) Proceedings of the 13th international conference on the practice and theory of automated timetabling - PATAT 2022, vol III, pp 208–210

Wolsey LA (2021) Integer programming, 2nd edn. Wiley, New York

Chapter 5
Case Studies

This chapter reports in detail three applications of sports tournament scheduling. They involve the use of approximate and exact methods described in the previous chapters.

Section 5.1 starts with a description of the problem of minimization of carry-over effects. Two real-life cases in which carry-over effects affected a tournament's balance or outcome are illustrated. The weighted variant of the problem is described, together with its integer programming formulation. An approximate algorithm hybridizing the ILS metaheuristic with a multistart strategy is presented, and computational results are discussed.

Section 5.2 reports on the formulation, implementation, and practical use of the optimization software developed for scheduling the first division of the annual Brazilian football tournament. Fair and balanced schedules for all teams are significant issues ensuring attractiveness and confidence in the tournament's outcome. We summarize the main characteristics of the tournament and discuss the requirements the schedule should meet, followed by its integer programming formulation. The solution approach is based on a three-phase decomposition strategy. We discuss the practical experience of using the software developed to schedule the tournament from 2009 to 2011.

Section 5.3 shows that electronic sports tournaments are adapted to dynamic scheduling. It describes a dynamic approach for scheduling tournaments based on a modification of the Swiss-system design. Colley's ranking method updates all competitors' ratings at every round. The ratings are used for maximizing game fairness and viewers' utility in the integer programming formulation of the team pairing and game scheduling problem solved in each round. The approach was validated using real-life data from the 2020 Honor of Kings World Champion Cup group stage and further evaluated using randomly generated test problems with up to 80 competitors.

The chapter concludes with a collection of real-life applications of sports scheduling.

C. C. Ribeiro et al., *Combinatorial Models for Scheduling Sports Tournaments*, EURO Advanced Tutorials on Operational Research,
https://doi.org/10.1007/978-3-031-37283-4_5

5.1 Minimizing Weighted Carry-Over Effects in Round-Robin Tournaments

The value of the carry-over effects is one of the various measures to assess the quality of a round-robin tournament schedule. We describe the weighted variant of the minimum carry-over effects problem, which is formulated by integer programming. An approximate algorithm based on the hybridization of the iterated local search metaheuristic with a multistart strategy is proposed, followed by the discussion of numerical results.

5.1.1 Motivation

There are many relevant aspects to be considered in the determination of the best schedule for a tournament. In some situations, one looks for a schedule minimizing the total distance traveled, as in the case of the traveling tournament problem or its mirrored variant, which is common to many tournaments in South America. Other problems attempt to minimize the number of breaks, i.e., the number of pairs of consecutive home games or away games. The original formulation of the annual Brazilian football tournament scheduling was a bi-criteria optimization problem. The first objective consisted in maximizing the number of games that open TV channels could broadcast (to increase the revenues from broadcast rights). In contrast, the second consisted of finding a balanced schedule with minimum home and away breaks (for fairness).

The minimization of the carry-over effects value, introduced in Sects. 1.3.5 and 2.13, is a fairness criterion leading to an even distribution of the sequence of games along the schedule.

Suppose that team (or athlete) A plays team (or athlete) C right after playing team (or athlete) B. If B is much stronger than the other competitors, then C may take some advantage over A in the game between them. This situation is due to the great effort A would have made in the previous game against B. This type of situation in which one of the teams (or athletes) might be benefited should be avoided or, at least, minimized.

We say that C receives a carry-over effect of one unit due to B if a team (or an athlete) A plays C just after its game against B. We consider a single round-robin tournament played by $2n$ teams. Figure 5.1a displays an array describing a hypothetical tournament schedule. Rows are associated with rounds, and columns correspond to teams. The entry at the intersection of the row associated with round i with the column corresponding to team j gives the team playing against team j in round i. For example, team A plays with team H in the first round. We may count the carry-over effects each team gives to every other in the tournament schedule and build the tournament's carry-over effects matrix. Each entry in the row corresponding to team i and the column corresponding to team j of the carry-over

	Teams							
Rounds	A	B	C	D	E	F	G	H
1	H	C	B	E	D	G	F	A
2	C	D	A	B	G	H	E	F
3	D	E	F	A	B	C	H	G
4	E	F	H	G	A	B	D	C
5	F	G	E	H	C	A	B	D
6	G	H	D	C	F	E	A	B
7	B	A	G	F	H	D	C	E

(a)

	Teams							
Teams	A	B	C	D	E	F	G	H
A	0	0	3	0	1	2	1	0
B	5	0	0	0	1	0	0	1
C	0	1	0	3	0	3	0	0
D	0	2	0	0	2	0	3	0
E	1	1	0	2	0	2	0	1
F	0	0	0	0	2	0	3	2
G	0	3	1	0	0	0	0	3
H	1	0	3	2	1	0	0	0

(b)

Fig. 5.1 (**a**) Schedule of a single round-robin tournament with eight teams and (**b**) its carry-over effects matrix

matrix indicates the carry-over effect team i gives to team j. Consider teams A and F. Team A plays with team H in round 1, which plays with team F in round 2. Also, team A plays with team C in round 2, which plays with team F in round 3. Therefore, team A gives team F a carry-over effect of two units. Figure 5.1b presents the underlying carry-over effects matrix associated with the schedule represented in Fig. 5.1a.

A fair tournament regarding the carry-over effects is one in which the latter are evenly distributed over the cells of the carry-over effects matrix. Furthermore, the total carry-over effects value is minimized.

In this section, the original problem of minimizing the carry-over effects is extended to the minimization of the total weighted carry-over effects. The weighted problem is formulated in the next section. Section 5.1.3 describes a heuristic based on the hybridization of the iterated local search metaheuristic with a multistart strategy for approximately solving the weighted problem. Numerical results are reported and discussed in Sect. 5.1.4.

5.1.2 Formulation of the Weighted Problem

We consider a compact single round-robin tournament played by $2n$ teams in $2n-1$ rounds.

We say that a team i gives a carry-over effect to team j if another team plays consecutively against teams i and j. In other words, another team plays with team i in round t and with team j in round $t+1$, for some $t = 1, \ldots, 2n-1$. The rounds are considered cyclically, i.e., round $2n-1$ is followed by the first round. If team i is very strong (resp. weak) and several teams play consecutively against teams i and j in this order, then team j may be favored (resp. handicapped) when compared with other teams. In an ideal schedule concerning carry-over effects, no teams i, j, x, y

should exist such that both teams x and y play with team j immediately after playing with team i.

The carry-over effects matrix $C = (c_{ij})$, with $i, j = 1, \ldots, 2n$, measures the tournament balance for this criterion. Each entry c_{ij} of this matrix counts the number of carry-over effects given by team i to team j. It can be seen that $c_{ii} = 0$ and $\sum_{j=1}^{2n} c_{ij} = \sum_{i=1}^{2n} c_{ij} = 2n - 1$ for every row $i = 1, \ldots, 2n$ and every column $j = 1, \ldots, 2n$. The quality of a schedule concerning the carry-overs is measured by the value $\sum_{i=1}^{2n} \sum_{j=1}^{2n} c_{ij}^2$. An ideal or balanced schedule is one in which the lower bound value $2n(2n - 1)$ is achieved, i.e., all non-diagonal elements of the corresponding matrix C are equal to one.

The original (unweighted) carry-over effects minimization problem does not consider any information concerning the relative strengths of the teams: all carry-over effects have the same unit weight. However, in real-life sports competitions, the organizers may have an estimate of how well a team should perform in the competition, considering, e.g., its tradition, its ranking, the quality of its players, and the support of its fans in home games.

Therefore, minimizing the carry-over effects would not guarantee a fair schedule. A broader approach consists of assigning a weight w_{ij} to every ordered pair (i, j) of teams, based on their relative strengths or handicaps, and minimizing the total weighted carry-over effects.

For every pair of teams $i, j = 1, \ldots, 2n$ (with $i \neq j$) and for every round $k = 1, \ldots, 2n-1$ (with rounds cyclically represented, i.e., round $2n-1$ is followed by the first round), we define the following binary variable:

$$z_{ijk} = \begin{cases} 1, & \text{if team } i \text{ meets team } j \text{ in round } k, \\ 0, & \text{otherwise.} \end{cases}$$

We also define the number of carry-over effects given by team i to team j as

$$v_{ij} = \sum_{\ell=1}^{2n} \sum_{k=1}^{2n-1} z_{\ell ik} \cdot z_{\ell j,k+1}, \tag{5.1}$$

for $i, j = 1, \ldots, 2n : i \neq j$; $v_{ij} = 0$ if $i = j$.

A non-linear integer programming formulation of the weighted carry-over effects value minimization problem is presented below:

$$\min \sum_{i=1}^{2n} \sum_{j=1}^{2n} w_{ij} \cdot v_{ij}^2, \tag{5.2}$$

subject to:

$$z_{ijk} = z_{jik}, \qquad i, j, k = 1, \ldots, 2n, \tag{5.3}$$

$$\sum_{i=1}^{2n} z_{ijk} = 1, \qquad j, k = 1, \ldots, 2n, \tag{5.4}$$

$$\sum_{k=1}^{2n-1} z_{ijk} = 1, \qquad i, j = 1, \ldots, 2n : i \neq j, \tag{5.5}$$

$$z_{iik} = 0, \qquad i, k = 1, \ldots, 2n, \tag{5.6}$$

$$z_{ij,2n} = z_{ij1}, \qquad i, j = 1, \ldots, 2n, \tag{5.7}$$

$$z_{ijk} \in \{0, 1\}, \qquad i, j, k = 1, \ldots, 2n. \tag{5.8}$$

The objective function (5.2) minimizes the weighted sum of carry-over effects. Constraints (5.3) ensure that variables $z_{ijk} = z_{jik}$ are the same or that the game between teams i and j is the same as that between teams j and i (there is a unique game between teams i and j in a single round-robin tournament). Constraints (5.4) force every team to play exactly once in each round of a compact tournament. Constraints (5.5) and (5.6) guarantee that each team plays exactly once with every other team. Constraints (5.7) enforce that the $2n$-th dummy round is equivalent to the first. Constraints (5.8) impose the binary requirements on the variables.

This formulation has $O(n^3)$ variables and defines a quadratic minimization problem. The linearization of the objective function (5.2) leads to a new formulation with $O(n^4)$ variables.

The solver CPLEX was applied to the solution of unweighted (i.e., $w_{ij} = 1$, for $i, j = 1, \ldots, 2n : i \neq j$) instances of this reformulation. Although the solver has been able to find the optimal solution for $2n = 4$ in a couple of seconds and $2n = 6$ in a few minutes, it took about three days of computation time to come up with the optimal solution for $2n = 8$. CPLEX could not solve instances with ten or more teams in reasonable times, motivating the use of heuristics for larger instances.

5.1.3 Hybrid Heuristic for Weighted Minimization

We describe a tailored heuristic for the minimization of the weighted carry-over effects. Algorithm 16 is based on the hybridization of the iterated local search metaheuristic introduced in Sect. 3.4.5 with a multistart strategy. Each iteration has two main steps: a multistart phase and an ILS phase. Each complete run comprises independent sequences of these two phases and returns the best solution found. Line 1 initializes the cost f^* of the best known solution. The loop in lines 2–23 is performed ten times. A starting schedule S_0 is built in line 3 in each iteration. The cost $f^*_{multistart}$ of the best solution found during the multistart phase is initialized in line 4. The multistart phase generates 100 new schedules. Each new schedule is obtained in line 6 by a constructive method from the starting schedule S_0. Local search is applied to this schedule in line 7. The best schedule obtained during the multistart phase is updated in lines 8–11. The starting solution S for the ILS phase is set as the best solution $S^*_{multistart}$ found during the multistart phase in line 13. The loop performs the ILS phase in lines 14–22, executed until a specific stopping criterion is satisfied. Each ILS iteration starts by applying a perturbation to the current solution S in line 15 and obtains the solution S'. Solution S' is tentatively improved by local search in line 16, and the resulting solution S'' is used to update the best known solution and its value in lines 17–20. Solution S'' obtained in line 16 replaces the current solution S in line 21 if some acceptance criterion is met, and a new ILS iteration resumes. The best solution found S^* is returned in line 23. The main steps of the algorithm's pseudo-code are described in more detail in the following sections.

Algorithm 16: Hybrid ILS with multistart heuristic for weighted carry-over

Output: Locally optimal solution S^*
1 $f^* \leftarrow \infty$;
2 **for** $i = 1$ ***to*** 10 **do**
3   Build a starting schedule S_0;
4     $f^*_{multistart} \leftarrow \infty$;
5   **repeat**
6   Create a new schedule S from S_0 with a constructive method;
7   Apply local search to schedule S to obtain a new schedule S';
8   **if** $f(S') < f^*_{multistart}$ **then**
9             $S^*_{multistartl} \leftarrow S'$;
10             $f^*_{multistart} \leftarrow f(S')$;
11   **end if**
12   **until** *100 schedules have been generated*;
13     $S \leftarrow S^*_{multistart}$;
14   **while** *stopping criterion not reached* **do**
15   Obtain a new schedule S' by applying a perturbation to S;
16   Apply local search to schedule S' to obtain a new schedule S'';
17   **if** $f(S'') < f^*$ **then**
18             $S^* \leftarrow S''$;
19             $f^* \leftarrow f(S'')$;
20   **end if**
21   Replace the current solution S by S'' using an acceptance criterion;
22   **end while**
23 **end for**
24 Return the best solution S^*.

5.1.3.1 Construction Method for Building New Schedules

Each application of the multistart construction phase builds a starting schedule S_0 in line 3 of Algorithm 16, using any of the two approaches described below.

The first approach is the circle method presented in Sect. 2.2, which gives the so-called canonical 1-factorization. It can be used for any value of $2n$. Assuming the teams are numbered from 1 to $2n$, the edge-set of the 1-factor corresponding to round $k = 1, \dots, 2n-1$ is given by

$$F_k = \{[k, 2n]\} \cup \{[\mathrm{mod}^{(2n-1)}(k+\ell), \mathrm{mod}^{(2n-1)}(k-\ell)] : \ell = 1, \dots, n-1\}, \tag{5.9}$$

with

$$\mathrm{mod}^{(2n-1)}(x) = \begin{cases} x + 2n - 1, & \text{if } x \leq 0, \\ x, & \text{if } 1 \leq x \leq 2n-1, \\ x - (2n-1), & \text{if } x \geq 2n. \end{cases} \tag{5.10}$$

The second approach can be applied whenever $2n$ is a multiple of four. It first separates the $2n$ teams in two sets $V_1 = \{1, \dots, n\}$ and $V_2 = \{n+1, \dots, 2n\}$. The first n rounds are made up only of games with one team in V_1 and the other in V_2. The games scheduled for round k correspond to the edges of the 1-factor whose edge set is defined as $\{(\ell, c(k, \ell)) : \ell = 1, \dots, n\}$, for $k = 1, \dots, n$, with

$$c(k, \ell) = \begin{cases} k + \ell + n - 1, & \text{if } (k + \ell - 2) < n, \\ k + \ell - 1, & \text{if } (k + \ell - 2) \geq n. \end{cases} \tag{5.11}$$

Each of the last $n-1$ rounds is formed by combining a 1-factor of a canonical 1-factorization of the complete graph associated with V_1 with a 1-factor of a canonical 1-factorization of the complete graph associated with V_2.

Line 6 creates a new schedule in every 100 repetitions of the repeat loop in lines 5–12. First, two rounds of the starting schedule S_0 are randomly selected and used as the first two rounds of the new solution. In each subsequent step, one new round of the starting schedule S_0 is selected to be used in the current, incomplete schedule. This process continues until a complete solution is obtained. The rules used for selecting and placing rounds of the starting schedule in the incomplete solution under construction mimic those in the well-known nearest neighbor heuristic and arbitrary insertion heuristics for the traveling salesman problem:

- *nearest neighbor heuristic*: select a new round that causes the minimum increment in the weighted carry-over effects and place it as the last in the solution under construction.
- *arbitrary insertion heuristic*: select randomly any new round and insert it between two rounds minimizing the increase in the weighted carry-over effects.

One of these strategies is randomly chosen with equal probability to generate a new schedule at each iteration of the multistart phase.

5.1.3.2 Local Search

The schedules built by the construction method described in the previous section in line 6 are tentatively improved by local search in line 7 of Algorithm 16. The local search procedure follows the variable neighborhood descent strategy described in Sect. 3.3.3. The best-improving move in each neighborhood is applied to the current solution. Neighborhoods team swap (TS), round swap (RS), partial round swap (PTS), and partial team (PTS) are used in this order, all of them previously presented in Sect. 3.5.

If the locally optimal solution (for all four neighborhoods) obtained by the variable neighborhood descent local search strategy improves the best schedule found in this multistart phase, then the latter is updated in lines 8–11.

5.1.3.3 Iterated Local Search

The best schedule obtained at the multistart phase is saved in line 13 of Algorithm 16 to be used as the starting solution S by the iterated local search phase in lines 14–22, until some stopping criterion is reached.

Each ILS iteration starts in line 15 by applying a perturbation to the current solution S. The perturbation consists of a sequence of random moves in the PTS neighborhood, implemented as the game rotation ejection chain initially proposed for the traveling tournament problem; see Sect. 3.5.7. A randomly selected game is moved from the round where it is currently played to another randomly selected round. This first change is followed by a sequence of modifications to recover a feasible solution according to the ejection chain mechanism.

A bias is introduced in the perturbation procedure to drive the move away from pure randomness. Although one can not know beforehand the effect of a complete ejection chain on the weighted carry-over effects of a solution, the increment (or decrement) implied by the first step of the chain can be easily evaluated. The bias consists simply of forcing the game with the smallest first increment.

The same local search procedure described in Sect. 5.1.3.2 is applied to the perturbed solution S' in line 16. The best known solution S^* is updated in lines 17–20. The current solution S is replaced by S'' in line 21 if the carry-over effects value of the latter is less than or equal to $(1 + b)$ times the carry-over effects value of the former. The value of parameter b doubles after every $4n$ iterations without any modification of the current solution. It is reset to its initial value whenever the current solution is updated. The ILS phase stops if line 14 detects that a maximum number of deteriorating moves have been accepted since the last time the best solution was updated.

5.1.3.4 Iterated Local Search with Destructive Perturbations

Another perturbation strategy is based on using destructive neighborhoods to generate new solutions. Given the current solution, the main underlying idea consists in destroying randomly selected parts of the schedule while keeping all the rest untouched. The partial solution made up of the untouched parts is given as input to an integer programming solver to obtain a new solution as the outcome of the perturbation, different from the current one. The subproblem solved consists simply in filling up the schedule to obtain a different feasible solution satisfying constraints (5.3)–(5.8) subject to the variable fixations.

Two rules for selecting the parts to be destroyed were devised and considered in the experiments, leading to two different types of neighborhoods described next.

Rule A: Row Destruction

Consider the single round-robin tournament represented in Fig. 5.1a. Each row corresponds to a round. The games occurring in a given round are indicated in the corresponding row. A destructive perturbation move in this neighborhood is obtained by erasing some rows and replacing them with different ones. Given the current schedule, we impose that some selected rounds (rows) of the current solution will not occur in the new solution. Figure 5.2 illustrates a perturbation move in this neighborhood. The shaded rows correspond to the rounds selected to be destroyed and modified. All other rounds remain unchanged. The games of any destroyed round can not appear together in any round of the new solution.

Let $\bar{z}$ be the current solution. We recall that $\bar{z}_{ijk} = 1$ if and only if team i meets team j in round k of the current schedule, $\bar{z}_{ijk} = 0$ otherwise, for every pair of teams $i, j = 1, \ldots, 2n$ (with $i \neq j$) and for every round $k = 1, \ldots, 2n$. Furthermore,

	Teams									
Rounds	A	B	C	D	E	F	G	H	I	J
1	B	A	I	H	G	J	E	D	C	F
2	G	F	E	J	C	B	A	I	H	D
3	I	H	G	F	J	D	C	B	A	E
4	J	C	B	G	I	H	D	F	E	A
5	D	I	J	A	F	E	H	G	B	C
6	E	D	H	B	A	I	J	C	F	G
7	H	G	F	E	D	C	B	A	J	I
8	F	E	D	C	B	A	I	J	G	H
9	C	J	A	I	H	G	F	E	D	B

(a)

	Teams									
Rounds	A	B	C	D	E	F	G	H	I	J
1	J	C	B	H	I	G	F	D	E	A
2	G	F	E	J	C	B	A	I	H	D
3	I	H	G	F	J	D	C	B	A	E
4	B	A	I	G	H	J	D	E	C	F
5	C	J	A	I	F	E	H	G	D	B
6	E	D	H	B	A	I	J	C	F	G
7	H	G	F	E	D	C	B	A	J	I
8	F	E	D	C	B	A	I	J	G	H
9	D	I	J	A	G	H	E	F	B	C

(b)

Fig. 5.2 Perturbation move using the row destruction neighborhood applied to rounds 1, 4, 5, and 9 (these rounds are rescheduled by solving a restricted integer programming problem). (**a**) Current solution. (**b**) Perturbed solution

let $D \subseteq \{1, \ldots, 2n\}$ be the set of rows to be destroyed in the current solution. Constraints (5.12) are added to the formulation to fix all rows not in D, while constraints (5.13) force the destruction of all rows in D and avoid their repetition in any slot of the new solution:

$$z_{ijk} = \bar{z}_{ijk}, \qquad k \in \{1, \ldots, 2n\} \setminus D, \quad i, j = 1, \ldots, 2n : i \neq j, \tag{5.12}$$

$$\sum_{i=1}^{2n} \sum_{\substack{j=1 \\ i \neq j \text{ and } \bar{z}_{ijk}=1}}^{2n} z_{ij\ell} \leq 2n - 2, \qquad k, \ell \in D. \tag{5.13}$$

A different solution can not be generated if one single row is destroyed. If two rows are destroyed, then only one feasible schedule can be generated, corresponding to swapping these two rows. Experimental results showed that destroying three rows often makes it impossible to rebuild a feasible schedule. Therefore, the computational experiments considered only row destruction perturbation moves of exactly four rows, since the computation times would increase enormously if more extensive moves were permitted.

Rule B: Column Destruction

Considering the timetable of the round-robin tournament represented in Fig. 5.1a, each of its columns corresponds to the sequence of games played by a team. A destructive perturbation move in this neighborhood is obtained by erasing some columns and replacing them with different ones. Given the current schedule, we impose that some selected teams (columns) of the current solution will not have the same sequence of opponents in the new solution. Figure 5.3 illustrates a perturbation move in this neighborhood. The dark-shaded columns correspond to the teams selected to have their game sequences destroyed and modified. All other columns remain unchanged, except the light-shaded cells that correspond to games involving the teams associated with the columns selected for destruction.

Let again $\bar{z}$ be the current solution, with $\bar{z}_{ijk} = 1$ if and only if team i plays against team j in round k of the current schedule, $\bar{z}_{ijk} = 0$ otherwise, for every pair of teams $i, j = 1, \ldots, 2n$ (with $i \neq j$) and for every round $k = 1, \ldots, 2n$. Furthermore, let $D \subseteq \{1, \ldots, 2n\}$ be the set of columns to be destroyed in the current solution. Constraints (5.14) are added to the formulation to fix all games of the columns not in D whose associated opponent is also not in D, while constraints (5.15) are added to the formulation to enforce the destruction of all columns in D:

$$z_{ijk} = \bar{z}_{ijk}, \qquad k = 1, \ldots, 2n, \quad i, j \in \{1, \ldots, 2n\} \setminus D : j \neq i, \tag{5.14}$$

$$z_{ijk} \leq 1 - \bar{z}_{ijk}, \qquad k = 1, \ldots, 2n, \quad i \in D, \quad j = 1, \ldots, 2n : j \neq i. \tag{5.15}$$

	Teams									
Rounds	A	B	C	D	E	F	G	H	I	J
1	F	E	D	C	B	A	J	I	H	G
2	B	A	I	H	G	J	E	D	C	F
3	I	D	J	B	F	E	H	G	A	C
4	E	J	H	G	A	I	D	C	F	B
5	G	F	E	I	C	B	A	J	D	H
6	J	H	G	F	I	D	C	B	E	A
7	C	I	A	J	H	G	F	E	B	D
8	D	C	B	A	J	H	I	F	G	E
9	H	G	F	E	D	C	B	A	J	I

(a)

	Teams									
Rounds	A	B	C	D	E	F	G	H	I	J
1	J	E	H	I	B	G	F	C	D	A
2	B	A	I	F	G	D	E	J	C	H
3	I	F	D	C	H	B	J	E	A	G
4	E	D	F	B	A	C	H	G	J	I
5	G	H	E	J	C	I	A	B	F	D
6	F	J	G	H	I	A	C	D	E	B
7	C	I	A	G	J	H	D	F	B	E
8	H	C	B	E	D	J	I	A	G	F
9	D	G	J	A	F	E	B	I	H	C

(b)

Fig. 5.3 Perturbation move using the column destruction neighborhood applied to columns D, F, H, and J (columns in dark grey are destroyed, and the games of the corresponding teams are rescheduled, while cells in light grey represent changes in the opponents of the teams associated with the dark grey columns). (**a**) Current solution. (**b**) Perturbed solution

As for the case of row destruction, the computational experiments addressed column destruction perturbation moves involving the destruction of exclusively four columns.

5.1.4 Computational Results

Each instance is defined by the number of teams $2n$ and by a matrix of weights w determining the carry-over effects. Four classes of weighted instances have been generated for the computational experiments:

- Random instances: weights are randomly generated in the interval $[1, 4n]$. Three matrices identified by letters A, B, and C have been generated for each value of $2n$.
- Linear instances: a strength in the interval $[1, 2n]$ is assigned to each team. For simplicity, the strength of team i is set to i. Each weight w_{ij} is defined as the absolute value of the difference between the strengths of teams i and j.
- Perturbed linear instances: each weight of a linear instance is increased by a perturbation, randomly generated in the interval $[-n, n]$. Three instances identified by letters A, B, and C are generated for each value of $2n$. Absolute values are taken whenever the perturbation leads to a negative weight.
- Real-life inspired instances: six instances are derived from six editions of the annual Brazilian football tournament. The strength of each team is given by the number of points it obtained in the previous year. Each weight w_{ij} is defined as the absolute value of the difference between the strengths of teams i and j.

Instances made up only of unit costs, equivalent to those of the original, unweighted problem, were also considered.

The strategy used to generate initial solutions often strongly influences the quality of the solutions obtained by a metaheuristic. Tuning experiments showed that the best initial solutions were found when the circle method and the method described in Sect. 5.1.3.1 were indistinctly employed to obtain 1-factorizations. The same experiments suggested using 200 as the maximum number of deteriorating moves, only one ejection chain move in each perturbation, and parameter $b = 0.01$.

5.1.4.1 Weighted Instances

The hybrid heuristic was run five times for each instance, with the parameter values and implementation strategies selected above. Average and best values of the carry-over effects and average and worst computation times in seconds over the five runs are reported for each class of the weighted test instances, together with the average deviation from the best value found.

Results for the random instances are shown in the upper part of Table 5.1. These instances present the shortest computation times over all classes. The ILS phase always improved the solutions found in the multistart phase, except for the three smallest instances.

We consider next the results for the linear instances, reported in the lower part of Table 5.1. The ILS phase improved the solutions found in the multistart phase for all instances with $2n \geq 8$. The computation times are much greater than those observed for the previous class of instances. This result is particularly observable for the problem instance with $2n = 20$ teams.

The upper part of Table 5.2 shows the results for the perturbed linear instances. As for the two previous classes, the ILS phase constantly improved the best solution found in the multistart phase, except for three small instances. The computation times are greater than those observed for the random and linear classes. For both classes of random and linear instances, the heuristic found high-quality solutions that are, on average, within 2% of the best values obtained, illustrating its robustness.

Finally, the results obtained for the real-life inspired instances are shown in the lower part of Table 5.2. This class presents the highest computation times for the instances with $2n = 20$ teams. This behavior seems to be due to the greater weight diversity observed in this class. As for the previous classes of instances, for both the perturbed linear instances and the real-life-inspired instances, the heuristic was able to find high-quality solutions that are, on average, within 2% of the best values obtained.

CPLEX solved exactly all instances with eight or fewer teams. For all of them, the proposed heuristic found an optimal solution in all runs. Larger instances with ten or more teams could not be solved in 3 days of computations by CPLEX. The lower bounds provided by the linear programming relaxation were weak and did not help the solver.

Table 5.1 Results for the random (upper part) and linear instances (lower part)

Instance	Avg. COEV	Dev. (%)	Best COEV	Avg. time (s)	Worst time (s)
inst4randomA	63.0	0.0	63	0.2	1.0
inst4randomB	53.0	0.0	53	0.2	1.0
inst4randomC	45.0	0.0	45	0.0	0.0
inst6randomA	233.0	0.0	233	0.8	1.0
inst6randomB	274.0	0.0	274	0.8	1.0
inst6randomC	235.0	0.0	235	1.2	2.0
inst8randomA	505.0	0.0	505	7.2	8.0
inst8randomB	495.0	0.0	495	18.6	20.0
inst8randomC	470.0	0.0	470	13.8	14.0
inst10randomA	912.6	2.0	895	139.4	150.0
inst10randomB	793.0	1.5	781	151.0	173.0
inst10randomC	744.6	1.0	737	149.0	172.0
inst12randomA	1549.2	1.9	1521	453.2	560.0
inst12randomB	1520.2	2.1	1489	402.4	451.0
inst12randomC	1594.4	1.4	1572	455.8	546.0
inst14randomA	2620.0	0.5	2608	39.8	44.0
inst14randomB	2903.8	1.2	2870	38.0	45.0
inst14randomC	2777.6	0.6	2760	36.8	42.0
inst16randomA	3812.2	1.5	3756	2605.8	2878.0
inst16randomB	3846.6	1.3	3797	2848.2	3294.0
inst16randomC	3773.6	0.5	3755	3138.0	3483.0
inst18randomA	5574.4	1.1	5515	1381.2	2442.0
inst18randomB	5816.2	1.2	5745	792.4	1271.0
inst18randomC	5598.0	0.9	5548	1547.2	3157.0
inst20randomA	7764.4	0.1	7760	136.8	145.0
inst20randomB	7928.0	0.5	7888	147.8	174.0
inst20randomC	7700.2	0.8	7636	136.8	144.0
Average		0.7			
inst4linear	20.0	0.0	20	0.2	1.0
inst6linear	114.0	0.0	114	0.6	1.0
inst8linear	168.0	0.0	168	8.4	9.0
inst10linear	318.0	0.0	318	64.6	70.0
inst12linear	504.0	1.6	496	271.4	309.0
inst14linear	960.0	0.2	958	18.6	20.0
inst16linear	1088.0	1.1	1076	2007.2	2313.0
inst18linear	1698.0	2.3	1660	4102.0	4729.0
inst20linear	2257.6	2.1	2212	5213.0	5624.0
Average		0.8			

Table 5.2 Results for perturbed linear instances (upper part) and real-life inspired instances (lower part)

Instance	Avg. COEV	Dev. (%)	Best COEV	Avg. time (s)	Worst time (s)
inst4perturbedlinearA	16.0	0.0	16	0.2	1.0
inst4perturbedlinearB	17.0	0.0	17	0.2	1.0
inst4perturbedlinearC	22.0	0.0	22	0.0	0.0
inst6perturbedlinearA	68.0	0.0	68	1.4	2.0
inst6perturbedlinearB	73.0	0.0	73	0.6	1.0
inst6perturbedlinearC	60.0	0.0	60	0.8	1.0
inst8perturbedlinearA	137.0	0.0	137	31.0	32.0
inst8perturbedlinearB	141.0	0.0	141	22.4	23.0
inst8perturbedlinearC	162.0	0.0	162	26.8	28.0
inst10perturbedlinearA	329.0	0.9	326	136.0	162.0
inst10perturbedlinearB	277.0	1.1	274	134.0	163.0
inst10perturbedlinearC	291.8	2.7	284	128.6	152.0
inst12perturbedlinearA	601.2	2.4	587	484.2	638.0
inst12perturbedlinearB	528.6	0.7	525	494.6	580.0
inst12perturbedlinearC	486.6	1.8	478	546.6	648.0
inst14perturbedlinearA	940.0	2.2	920	42.8	45.0
inst14perturbedlinearB	947.4	1.7	932	39.8	43.0
inst14perturbedlinearC	993.2	0.3	990	36.2	40.0
inst16perturbedlinearA	1403.0	2.0	1376	3432.8	3905.0
inst16perturbedlinearB	1360.2	0.9	1348	3343.8	3668.0
inst16perturbedlinearC	1118.0	1.8	1098	3961.0	4683.0
inst18perturbedlinearA	2063.4	2.9	2005	5162.4	6983.0
inst18perturbedlinearB	1965.4	2.3	1921	4038.0	5343.0
inst18perturbedlinearC	1712.0	8.0	1585	4833.2	5805.0
inst20perturbedlinearA	3092.2	0.9	3065	482.0	1304.0
inst20perturbedlinearB	2866.8	2.4	2800	2923.4	4719.0
inst20perturbedlinearC	2837.2	1.7	2791	1479.6	3509.0
Average		1.4			
inst24brasileirao2003	7730.4	2.5	7542	13897.8	15755.0
inst24brasileirao2004	7179.6	1.3	7088	12992.8	13468.0
inst22brasileirao2005	5228.8	1.4	5158	10599.0	13959.0
inst20brasileirao2006	5310.0	1.4	5236	5705.4	6358.0
inst20brasileirao2007	4876.0	0.9	4834	2715.8	3655.0
inst20brasileirao2008	4045.6	2.6	3944	6805.6	8308.0
Average		1.7			

5.1.4.2 Validation: Unweighted Instances

To validate the proposed hybrid heuristic, we also applied it to instances of the original unweighted problem. These instances are equivalent to weighted instances with unit weights.

The heuristic matched the best known results for $2n = 4, 6, 8, 10$, and 16. Although it could not match the best known solution for $2n = 14$, it improved the best known upper bound for $2n = 12$ by almost 10%, producing a previously unknown solution whose value of the carry-over effects is only 160.

The ILS phase of the heuristic terminates after 200 deteriorating moves have been accepted without improvement in the best solution found. For all classes of weighted instances, we observed that this counter very rarely reached values greater than 50. However, a very different behavior was noticed for the unweighted instances, for which many improvements obtained during the ILS phase were achieved after more than 50 deteriorating moves.

5.1.4.3 Destructive Neighboorhoods

This section reports the computational results obtained using destructive perturbations and compares them with those presented in the previous section.

Table 5.3 displays the results obtained with the three types of perturbations for the unweighted instances. It also displays the results obtained when one of the three neighborhoods was randomly selected with uniform probability to be used at each

Table 5.3 Results for different perturbations: game rotation, row destruction, column destruction, and random neighborhood selection

Perturbation	$2n$	Avg. COEV	Best COEV	Avg. time	Longest time (s)
Game rotation	10	109.2	108.0	7.2	8.0
	12	164.4	160.0	15.2	19.0
	14	254.0	254.0	5.6	6.0
	16	244.8	240.0	71.2	83.0
	Averages	193.1	190.5	24.8	29.0
Row destruction	10	108.0	108.0	311.8	330.0
	12	160.0	160.0	894.6	1068.0
	14	246.0	244.0	1093.2	1220.0
	16	240.0	240.0	1700.2	1913.0
	Averages	188.5	188.0	1000.0	1132.8
Column destruction	10	108.0	108.0	60.6	68.0
	12	161.6	160.0	144.2	158.0
	14	248.0	246.0	269.0	298.0
	16	240.0	240.0	566.6	785.0
	Averages	189.4	188.5	260.1	327.3
Random selection	10	108.0	108.0	31.8	35.0
	12	162.4	160.0	72.8	78.0
	14	248.8	246.0	96.6	113.0
	16	243.2	240.0	309.6	386.0
	Averages	190.6	188.5	127.7	153.0

iteration. For each value of $2n = 10$, 12, 14, and 16, we report the average and best carry-over effects values and the average and worst computation times (in seconds) over the five runs of the heuristic.

The game rotation perturbations provided the fastest (by far) algorithm. However, the destructive perturbations improved the value of the carry-over effects for the problem instance with $2n = 14$ teams and matched the values obtained for the other instances. The algorithm using row destruction perturbations always found the best values over all variants, although at the cost of a substantial computational effort. The use of column destruction perturbations, on the other hand, demanded much less computational effort while still providing better results than the game rotation perturbations. Randomly selecting the neighborhood used at each iteration is a good compromise.

We observe that the column destruction neighborhoods are much less time-consuming than employing row destruction because the first fixes fewer elements of the schedule than the second, making it easier for the solver to find a feasible solution to replace the current one.

5.2 Scheduling the Annual Brazilian Football Tournament

The annual football tournament organized by the Brazilian Football Confederation (CBF) is Brazil's most important sports event. Its primary sponsor used to be TV Globo, Brazil's largest media group and television network.

Fair and balanced schedules for all teams are significant issues ensuring attractiveness and confidence in the tournament's outcome. Furthermore, the TV sponsors expect schedules that allow open channels to broadcast as many important games as possible. Big cities hosting two or more teams and a large number of fans impose additional security constraints to avoid clashes of fans before or after the games.

This section reports on the formulation, implementation, and practical use of the optimization software developed in partnership with CBF to determine good schedules for the annual Brazilian football tournament's first division (Series A). Section 5.2.1 summarizes the main characteristics of the tournament, including its structure and the participating teams. Section 5.2.2 discusses the requirements that the schedule should satisfy. The associated integer programming formulation is presented in Sect. 5.2.3. Section 5.2.4 describes the solution approach based on a three-phase first-break, then-schedule strategy. Section 5.2.5 presents some statistics related to software development and algorithm performance and reports on the practical experience resulting from using the proposed system in real-life to schedule the editions of 2009, 2010, and 2011 of the tournament.

5.2.1 Tournament Structure

The tournament typically lasts 7 months, from April to December, and is organized in two main divisions (Series A and B). Each division has $2n$ teams and follows a compact mirrored double round-robin design. Each team has its venue in its home city. Each game is played at the venue of one of the two teams in confrontation.

Weekend rounds are played on Saturdays and Sundays, while *midweek rounds* are played on Wednesdays and Thursdays. The dates available for game playing change from year to year and are coordinated with other competitions, such as the state tournaments, Brazil's Cup, South America's Cup, and CONMEBOL Libertadores. Since the tournament is compact, each phase has exactly $2n - 1$ rounds. For any $k = 1, \dots, 2n - 1$ such that round k is a weekend round in both tournament phases, we say that it is a *double weekend round*. Some specific games are required to be played in double weekend rounds.

Each year, the teams ranked in the last four positions of the standings table in Series A are downgraded to play Series B in the next year. In contrast, the four best-ranked teams in Series B are upgraded to play Series A. TV rights, marketing revenues, and gate attendance are much higher for Series A games and teams than those in Series B.

Twelve teams belong to the so-called "Group of Twelve" (G12), formed by the league's most traditional or founding teams. They usually have more considerable broadcast rights. Not all of them necessarily play in Series A every year.

Teams playing in the same division are organized in pairs (determined by CBF), whose teams have complementary profiles of game playing (home or away). Usually, teams in the same pair are based in the same home city. Table 5.4 displays the 20 teams participating in the 2010 edition of the tournament, their home cities, the states in which their home cities are located, and an indication of whether they belong to the G12. Santos, an important city with a well-known team, is very close to São Paulo. For the sake of simplicity, Santos is also considered one of the teams of the city of São Paulo.

São Paulo is the wealthiest state in Brazil and has several strong teams in its capital and also in smaller, economically dynamic cities. In contrast, the teams from all other states have their home cities exclusively in the state capitals.

This section concludes with two additional definitions. *Regional games* involve two opposing teams whose home cities are located in the same state. *Classic games* (or *derbies*) are those that involve two opponents based in the same home city, that is, teams based in Rio de Janeiro, São Paulo (including Santos), Porto Alegre, Belo Horizonte, or Goiânia, for the 2010 edition of the tournament. Classic games involve teams with a long tradition of rivalry. They are usually the most important games and attract more attendees, fans, and media interest. All classic games are also regional games.

Table 5.4 Participating teams in Series A of the 2010 edition of the Brazilian football tournament, including their home cities, home states, and an indication of whether a team belongs to G12 (teams in the same pair appear in consecutive rows and are determined by CBF; teams in odd rows play at home in the tournament's first round)

Index	Team	Home city	State	G12?
1	Atlético Goianiense	Goiânia	Goiás	No
2	Goiás	Goiânia	Goiás	No
3	Atlético Mineiro	Belo Horizonte	Minas Gerais	Yes
4	Cruzeiro	Belo Horizonte	Minas Gerais	Yes
5	Avaí	Florianópolis	Santa Catarina	No
6	Atlético Paranaense	Curitiba	Paraná	No
7	Botafogo	Rio de Janeiro	Rio de Janeiro	Yes
8	Fluminense	Rio de Janeiro	Rio de Janeiro	Yes
9	Ceará	Fortaleza	Ceará	No
10	Vitória	Salvador	Bahia	No
11	Corinthians	São Paulo	São Paulo	Yes
12	São Paulo	São Paulo	São Paulo	Yes
13	Flamengo	Rio de Janeiro	Rio de Janeiro	Yes
14	Vasco da Gama	Rio de Janeiro	Rio de Janeiro	Yes
15	Internacional	Porto Alegre	Rio Grande do Sul	Yes
16	Grêmio	Porto Alegre	Rio Grande do Sul	Yes
17	Guarani	Campinas	São Paulo	No
18	Grêmio Prudente	Presidente Prudente	São Paulo	No
19	Palmeiras	São Paulo	São Paulo	Yes
20	Santos	Santos	São Paulo	Yes

5.2.2 Schedule Requirements

The tournament schedule must satisfy many constraints, ranging from fairness to security issues and from technical to broadcasting criteria. Some of them reflect strategies for maximizing revenues and tournament attractiveness; others attempt to avoid unfair situations that could benefit one team by scheduling a more convenient sequence of games. Teams, federations, city administrators, security agencies, and sponsors have discussed and established these requirements over the years. They can be slightly modified each year to accommodate specific situations. The constraints of the 2010 edition of the tournament can be grouped into the following five classes:

5.2.2.1 Round-Robin Constraints

- (A.1) Each team must meet every other twice, once at home and once away (double round-robin tournament).
- (A.2) Each team must play exactly once in each round, either home or away (compact schedule).

- (A.3) Each team must meet every other exactly once in the first (resp. second) phase along the $2n - 1$ initial (resp. last) rounds. Games in the second phase are played precisely in the same order as in the first but with interchanged venues. Consequently, the schedule of the second phase is directly determined by that of the first phase (i.e., a mirrored schedule).

5.2.2.2 Home-Away Profiles of Game Playing

- (B.1) Teams in the same pair always have complementary profiles of game playing: whenever one plays at home, the other plays away. Whenever a game is played between two teams from the same home city, only one acts as the home team with home rights.
- (B.2) Teams playing at home in the first round are determined by CBF, depending on historical data (teams that played away in the first round of the previous year's tournament and teams upgraded from Series B should play at home in the first round).
- (B.3) Teams alternate their playing condition (home or away) in every round during the first four rounds of the first phase and during the last two rounds of the second phase (i.e., there are no breaks in the first four rounds or in the last round of the tournament).
- (B.4) Teams playing at home in the first round necessarily play away in the last, and vice-versa.
- (B.5) All teams have the same number of home breaks and away breaks in each of the two phases of the tournament.
- (B.6) The number of breaks should be kept as low as possible. Section 5.2.4 showed that feasible schedules with a minimum number of breaks must have one home break and one away break for each team in each tournament phase.
- (B.7) Not counting the game with the other team in its pair, each team plays nine games at home and nine games away in each phase of the schedule.

5.2.2.3 Classic and Regional Games

- (C.1) As many classic games as possible should be played in double weekend rounds. Fans and the general public have more free time on weekends to spend on leisure activities, such as attending stadium games or watching them on TV. This requirement enables larger attendances and audiences for classic games, which are often characterized by historic local rivalries. It will be handled by the objective function of the formulation presented in Sect. 5.2.3.
- (C.2) No regional games can be played during the first three rounds (because the fans are less motivated) or in the last four rounds (to avoid travel schedules that are advantageous or disadvantageous to a team in a city that hosts multiple teams) of the tournament.

- (C.3) Whenever a classic game is played at São Paulo in some round, there are no games between G12 teams of Rio de Janeiro and São Paulo in this same round; this avoids competition that could divide the interest of the public.
- (C.4) Whenever a classic game is played at Rio de Janeiro in some round, there are no games between G12 teams of Rio de Janeiro and São Paulo in this same round; this avoids competition that could divide the interest of the public.
- (C.5) There can be at most one classic game played in each city in any round. This constraint is imposed because of the logistics necessary to avoid security and circulation problems.
- (C.6) No team can play two classic games in consecutive rounds because they are usually more demanding games for which the teams request more interplay time. Losing two classic games in a row could strongly affect a team's motivation and the interest of its fans.
- (C.7) Some games may have to be scheduled at predefined given rounds, according to the wishes of teams, sponsors, and organizers.

5.2.2.4 Geographical and G12 Constraints

- (D.1) No team whose home city is located in the state of São Paulo can play five or more consecutive games in that state. The other teams would consider it an advantage because such cities are geographically close.
- (D.2) No team whose home city is located in a state other than São Paulo can play four or more consecutive games in that state. This requirement avoids the advantage of a team staying at home for an extended period while other teams are traveling across the country.
- (D.3) Each team must play a game outside the state in which its home city is located in one of the first two rounds. This requirement is automatically implied by (B.3) and (C.2).
- (D.4) There are at least two and at most four games involving only G12 teams in every round; this generates an even distribution of games between strong teams throughout the tournament. It also ensures the existence of multiple games between strong teams in any round, offering more choices for broadcasting by TV channels.
- (D.5) A team can play at most five consecutive games against G12 teams to avoid a long series of challenging games.

5.2.2.5 Perfect Matching of Paired Teams

- (E.1) In the quest for a fair and balanced schedule, CBF imposes additional constraints to enforce a tight equilibrium to any two teams belonging to the same pair. Let A and B be two teams belonging to the same pair and C and D two other teams belonging to any other pair. If A plays with C at home (resp. away) in the first phase, then it plays away (resp. at home) with D in this phase. Consequently,

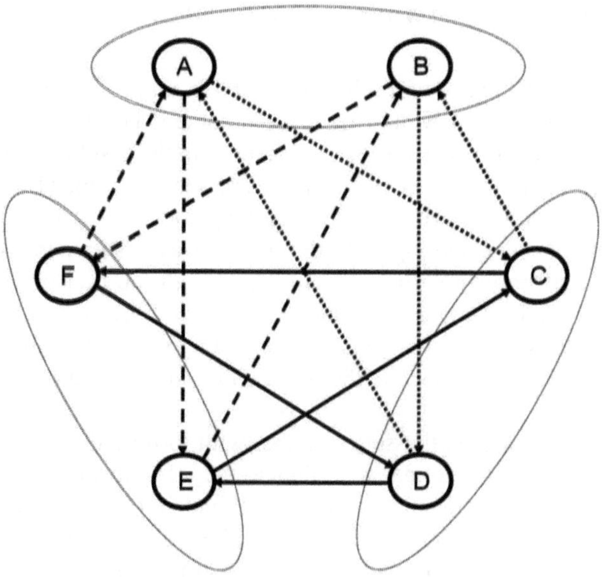

Fig. 5.4 Perfect matching of six teams paired into three pairs (an arrow in the arc from team i to team j means that this game is played at the venue of team j)

in this phase, B will play away (resp. at home) with C, and at home (resp. away) with D. The same constraints are automatically implied for the second phase since all its games have interchanged venues with respect to those in the first phase. Such constraints lead to an almost perfect equilibrium between teams in the same cities and regions and are considered among the most important to be enforced by CBF officials. Figure 5.4 illustrates a possible perfect matching of six paired teams, which applies to every subset of four teams organized in two pairs.

The maximization of gate attendance and TV audiences is the central issue in the schedule of the Brazilian tournament. Most revenues earned by the teams come from broadcast and merchandising rights paid by the sponsors, who request good schedules that draw large audiences. Fair and balanced schedules are also a significant issue for the attractiveness of the tournament and the confidence in its outcome, and they play a significant role in the competition's success. Consequently, determining a good schedule is mainly an issue of finding a solution that maximizes gate attendance and TV audiences while satisfying requirements that impose intricate patterns of game-playing that are hard to compute. To maximize gate attendance and TV audiences, we seek a schedule with a maximum number of classic games played in double weekend rounds. Most of the constraints handle fairness and equilibrium requirements.

5.2.3 Integer Programming Formulation

We recall that the first phase of a mirrored double round-robin tournament entirely determines the complete schedule of the tournament because the games in the second phase are played in the same order as those in the first but with interchanged venues. Therefore, the profiles of the teams can be seen as formed by two symmetric halves, with the second entirely determined by the first. An optimal schedule may be obtained by considering only the home-away assignments corresponding to the single round-robin tournament associated with the first phase. The integer programming formulation and the solution approach discussed in this section use this property because all assignments corresponding to games in the second phase are directly implied by those in the first. This strategy automatically handles requirements (A.3).

Let $T = \{1, \ldots, 2n\}$ be the set of teams, with those indexed by odd numbers playing at home in the first round, according to the recommendations made by CBF via requirement (B.2). Teams indexed by $2e - 1$ and $2e$ belong to the same pair, for $e = 1, \ldots, n$. Furthermore, let $H = \{1, \ldots, N\}$ be the set of all feasible profiles for the first phase of the tournament that satisfy constraints (B.3)–(B.6) and start with a home game. For every profile $\ell = 1, \ldots, N$ and every round $k = 1, \ldots, 2n - 1$, let $h(\ell, k) = 1$ if the team associated with the feasible profile indexed by ℓ plays at home in round k; $h(\ell, k) = 0$ otherwise.

We define the following decision variables:

$$x_{ijk} = \begin{cases} 1, & \text{if team } i \in T \text{ meets team } j \in T \text{ at home in round } k = 1, \ldots, 2n - 1, \\ 0, & \text{otherwise;} \end{cases}$$

and

$$y_{e\ell} = \begin{cases} 1, & \text{if team } 2e - 1 \text{ follows profile } \ell \in H \text{ and team } 2e \text{ follows its} \\ & \quad \text{complement, for every pair } e = 1, \ldots, n, \\ 0, & \text{otherwise.} \end{cases}$$

Let C be the set of all cities hosting two or more teams. Let $T(c) \subseteq T$ be the set of teams whose home city is $c \in C$. São Paulo is indexed by 1, and Rio de Janeiro by 2. Belo Horizonte is indexed by 3, Porto Alegre by 4, and Goiânia by 5. We also define SP as the set of teams whose home cities are located in the state of São Paulo. Let $\bar{R} \subset \{1, \ldots, 2n - 1\}$ be the set of rounds in which regional games can not be played according to requirement (C.2). In addition, let $D \subset \{1, \ldots, 2n-1\} \setminus \bar{R}$ be the set of double weekend rounds in which classic games should be played (whenever possible). We denote by $G12 \subset \{1, \ldots, 2n\}$ the subset of teams of G12 that play in the current edition of the tournament.

Thus, $C = \{$São Paulo, Rio de Janeiro, Belo Horizonte, Porto Alegre, Goiânia$\}$, $T(1) = \{11, 12, 19, 20\}$, $T(2) = \{7, 8, 13, 14\}$, $T(3) = \{3, 4\}$, $T(4) = \{15, 16\}$,

$T(5) = \{1, 2\}$, $SP = T(1) \cup \{17, 18\}$, $\bar{R} = \{1, 2, 3, 16, 17, 18, 19\}$, $D = \{11, 12, 13, 15\}$, and $G12 = \{3, 4, 7, 8, 11, 12, 13, 14, 15, 16, 19, 20\}$ for the 2010 edition of the tournament (see Table 5.4). We recall that except for the state of São Paulo, which has several teams in different cities, all teams are based in the other states' capitals.

The objective function (5.16) maximizes the number of classic games that can be played in double weekend rounds, leading to larger gate attendances and TV audiences as determined by requirement (C.1):

$$\max \sum_{k \in D} \sum_{c \in C} \sum_{i \in T(c)} \sum_{\substack{j \in T(c) \\ j \neq i}} (x_{ijk} + x_{jik}). \tag{5.16}$$

Constraints (5.17) force each team to meet every other team exactly once in each phase of the tournament, corresponding to requirement (A.1):

$$\sum_{k=1}^{2n-1} (x_{ijk} + x_{jik}) = 1, \qquad \forall i, j \in T : i < j. \tag{5.17}$$

Because requirement (B.6) imposes that each team must have one home break and one away break in each phase, teams who play at home in the first round will play ten home games and nine away games in the first phase of the tournament. Constraints (5.18) guarantee that the game between the two teams in each pair will be played at the home of the odd-indexed team in the tournament's first phase. Consequently, it will be played in the other team's home in the second phase. Therefore, each team in the pair will play nine games at home and nine games away against the other teams, following the requirement (B.7):

$$\sum_{k=1}^{2n-1} x_{2e-1,2e,k} = 1, \qquad e = 1, \ldots, n. \tag{5.18}$$

Constraints (5.19) express requirement (C.2):

$$\sum_{\substack{c \in C \\ c \neq 1}} \sum_{i \in T(c)} \sum_{\substack{j \in T(c) \\ j \neq i}} (x_{ijk} + x_{jik}) + \sum_{i \in SP} \sum_{\substack{j \in SP \\ j \neq i}} (x_{ijk} + x_{jik}) = 0, \qquad \forall k \in \bar{R}. \tag{5.19}$$

Constraints (5.20) (resp. (5.21)) handle requirements (C.3), (C.4), and (C.5), ensuring that there is at most one classic game played in São Paulo (resp. Rio de Janeiro) in any round. Moreover, whenever there is one such classic game, there are no games between G12 teams of Rio de Janeiro and São Paulo in the same round:

$$\sum_{i \in T(1)} \sum_{j \in T(2)} (x_{ijk} + x_{jik}) + 4 \cdot \sum_{i \in T(1)} \sum_{\substack{j \in T(1) \\ j \neq i}} (x_{ijk} + x_{jik}) \leq 4, \quad k = 1, \ldots, 2n-1, \tag{5.20}$$

$$\sum_{i \in T(2)} \sum_{j \in T(1)} (x_{ijk} + x_{jik}) + 4 \cdot \sum_{i \in T(2)} \sum_{\substack{j \in T(2) \\ j \neq i}} (x_{ijk} + x_{jik}) \leq 4, \quad k = 1, \ldots, 2n-1. \tag{5.21}$$

For any city hosting three or more teams, constraints (5.22) ensure that there is at least one round in between two classic games involving the same team, as imposed by requirement (C.6):

$$\sum_{\substack{i \in T(c) \\ i \neq j}} (x_{ijk} + x_{jik} + x_{ij,k+1} + x_{ji,k+1}) \leq 1,$$
$$\forall c \in C : |T(c)| \geq 3, \quad \forall j \in T(c), \quad k = 1, \ldots, 2n-2. \tag{5.22}$$

Constraints (5.23) and (5.24) enforce that any team whose home city is located in the state of São Paulo will play away at least once outside the state in any sequence of five consecutive games. Constraints (5.23) enforce requirement (D.1) for the first phase, while constraints (5.24) apply to the second phase:

$$\sum_{i \notin SP} (x_{ijk} + x_{ij,k+1} + x_{ij,k+2} + x_{ij,k+3} + x_{ij,k+4}) \geq 1,$$
$$\forall j \in SP, \quad k = 1, \ldots, 2n-5, \tag{5.23}$$

$$\sum_{i \notin SP} (x_{jik} + x_{ji,k+1} + x_{ji,k+2} + x_{ji,k+3} + x_{ji,k+4}) \geq 1,$$
$$\forall j \in SP, \quad k = 1, \ldots, 2n-5. \tag{5.24}$$

We note that constraints (5.19) that enforce requirement (C.2) also deal with requirement (D.1) for consecutive games in-between the first and second phases.

Similarly, constraints (5.25) and (5.26) enforce that any team whose home city is not located in the state of São Paulo will play away at least once outside the state where it is located in any sequence of four consecutive games. Constraints (5.25) enforce requirement (D.2) for the first phase, while constraints (5.26) apply to the second phase:

$$\sum_{i \notin T(c)} (x_{ijk} + x_{ij,k+1} + x_{ij,k+2} + x_{ij,k+3}) \geq 1,$$
$$\forall c \in C \setminus \{1\}, \quad \forall j \in T(c), \quad k = 1, \ldots, 2n-4, \tag{5.25}$$

$$\sum_{i \notin T(c)} (x_{jik} + x_{ji,k+1} + x_{ji,k+2} + x_{ji,k+3}) \geq 1,$$
$$\forall c \in C \setminus \{1\}, \quad \forall j \in T(c), \quad k = 1, \ldots, 2n - 4. \qquad (5.26)$$

Once again, we note that constraints (5.19) that enforce requirement (C.2) also deal with requirement (D.2) for consecutive games in-between the first and second phases.

Constraints (5.27) state that there are at least two and at most four games involving only G12 teams in every round, as determined by requirement (D.4):

$$2 \leq \sum_{i \in G12} \sum_{\substack{j \in G12 \\ j \neq i}} x_{ijk} \leq 4, \qquad k = 1, \ldots, 2n - 1. \qquad (5.27)$$

Constraints (5.28) enforce requirement (D.5), imposing that no team can play more than five consecutive games against G12 teams:

$$\sum_{\substack{j \notin G12 \\ j \neq i}} (x_{ij,k} + x_{ij,f(k+1)} + x_{ij,f(k+2)} + x_{ij,f(k+3)} + x_{ij,f(k+4)} + x_{ij,f(k+5)} +$$
$$x_{ji,k} + x_{ji,f(k+1)} + x_{ji,f(k+2)} + x_{ji,f(k+3)} + x_{ji,f(k+4)} + x_{ji,f(k+5)}) \geq 1,$$
$$\forall i \in T, \quad k = 1, \ldots, 2n - 1, \qquad (5.28)$$

where the function

$$f(a) = \begin{cases} a, & \text{if } a \leq 2n - 1, \\ a - (2n - 1), & \text{otherwise,} \end{cases}$$

is used to handle sequences of consecutive games that start in the first phase and finish in the second.

Constraints (5.29) and (5.30) enforce that each odd indexed team follows a different profile starting with a home game:

$$\sum_{\ell \in H} y_{e\ell} = 1, \qquad e = 1, \ldots, n, \qquad (5.29)$$

$$\sum_{e=1}^{n} y_{e\ell} \leq 1, \qquad \forall \ell \in H. \qquad (5.30)$$

Constraints (5.31) and (5.32) enforce that each odd indexed team follows the profile assigned to it and imply the satisfaction of requirement (A.2) for such teams:

$$\sum_{\substack{i \in T \\ i \neq 2e-1}} x_{2e-1,ik} = \sum_{\ell \in H} h(\ell, k) \cdot y_{e\ell}, \qquad \forall e = 1, \ldots, n, \quad k = 1, \ldots, 2n-1, \tag{5.31}$$

$$\sum_{\substack{i \in T \\ i \neq 2e-1}} x_{i,2e-1,k} = 1 - \sum_{\ell \in H} h(\ell, k) \cdot y_{e\ell}, \qquad \forall e = 1, \ldots, n, \quad k = 1, \ldots, 2n-1. \tag{5.32}$$

Similarly, constraints (5.33) and (5.34) determine whether each even indexed team plays at home or away in each round. Furthermore, they imply the satisfaction of requirements (A.2) for such teams:

$$\sum_{\substack{i \in T \\ i \neq 2e}} x_{2e,ik} = 1 - \sum_{\ell \in H} h(\ell, k) \cdot y_{e\ell}, \qquad \forall e = 1, \ldots, n, \quad k = 1, \ldots, 2n-1, \tag{5.33}$$

$$\sum_{\substack{i \in T \\ i \neq 2e}} x_{i,2e,k} = \sum_{\ell \in H} h(\ell, k) \cdot y_{e\ell}, \qquad \forall e = 1, \ldots, n, \quad k = 1, \ldots, n-1. \tag{5.34}$$

The complementarity of the profiles of teams in the same pair is also ensured by constraints (5.31)–(5.34), corresponding to requirement (B.1).

Finally, constraints (5.35) handle the perfect matching of paired teams imposed by requirement (E.1) by ensuring that any team i will play at home with precisely one of the two teams indexed by $2e - 1$ and $2e$, for $e = 1, \ldots, n$:

$$\sum_{k=1}^{2n-1} (x_{i,2e-1,k} + x_{i,2e,k}) = 1, \qquad e = 1, \ldots, n, \quad \forall i \in T : i \neq 2e-1, i \neq 2e. \tag{5.35}$$

The integrality constraints (5.36) and (5.37) complete the problem formulation:

$$x_{ijk} \in \{0, 1\}, \qquad \forall i \in T, \quad \forall j \in T : j \neq i, \quad k = 1, \ldots, 2n-1, \tag{5.36}$$

$$y_{e\ell} \in \{0, 1\}, \qquad e = 1, \ldots, n, \quad \ell = 1, \ldots, N. \tag{5.37}$$

5.2.4 Solution Approach

The problem has $O(n^3 + n|H|)$ variables and $O(n^2 + |H|)$ constraints. In Sect. 5.2.4.1, we show that the number $|H|$ of profiles with exactly two breaks is bounded by $O(n^2)$. Therefore, the full formulation has $O(n^3)$ variables and $O(n^2)$ constraints. The complete integer programming formulation defined by the objective function (5.16) and constraints (5.17)–(5.37), enumerating all N candidate profiles starting with a home game, could not be solved by a commercial solver such as CPLEX within an entire day of computations.

A three-phase solution approach was developed based on a first-break, then-schedule decomposition scheme. In the first phase, feasible profiles are created. In the second phase, a different feasible profile is assigned to each team because different teams must have different profiles in every feasible round-robin schedule. Finally, in the third phase, the approach attempts to find an optimal schedule by solving a simpler version of the integer programming formulation presented in Sect. 5.2.3, derived from the latter by variable fixations. In this simpler formulation, the binary variables $y_{e\ell}$ no longer exist because each team has already been assigned to a previously determined profile. Figure 5.5 illustrates the approach, whose three phases are described in the sections below.

Before performing the first phase of the decomposition solution approach, the algorithm computes an upper bound to the maximum number of classic games that can be played on double weekend rounds. Three classic games can be played easily in any double weekend round: Atlético Mineiro vs. Cruzeiro, Grêmio vs. Internacional, and Goiás vs. Atlético Goianiense. Because four teams have São Paulo as their home city, they play six games between them in each tournament phase. If six or more double weekend rounds are available for game playing, all such six games may be played in double weekend rounds. Otherwise, the number of games that can be played in double weekend rounds is bounded by the number of double weekend rounds. The same reasoning applies to Rio de Janeiro, the home city of four teams. For example, we consider the 2010 calendar, in which only four rounds were assigned to weekends in both tournament phases: 11, 12, 13, and 15. Therefore, at most, four of the six classic games to be played in São Paulo can take place in double weekend rounds. Suppose three classic games are played in three consecutive rounds in any of these two cities. In that case, at least one team will play two consecutive classic games, which is forbidden by requirement (C.6). Therefore, only three classic games can be played in São Paulo (and Rio de Janeiro) in double weekend rounds. Consequently, only 9 of 15 classic games can be played in double weekend rounds. This number establishes an upper bound to the value of the optimal solution: three classic games in the three cities that host two teams each, three in São Paulo and three in Rio de Janeiro.

5.2.4.1 Phase 1: Profile Generation

Profiles of mirrored schedules may be divided into two symmetric halves. The second half is entirely determined by the first. Therefore, we may determine which properties the first half of a profile must satisfy to make the entire profile feasible. If the number of breaks in the first half of a profile is even, then the total number of breaks is even and equal to twice the number of breaks in the first half. However, if the number of breaks in the first half is odd, then the total number of breaks is also odd and equal to twice the number of breaks in the first half plus one break

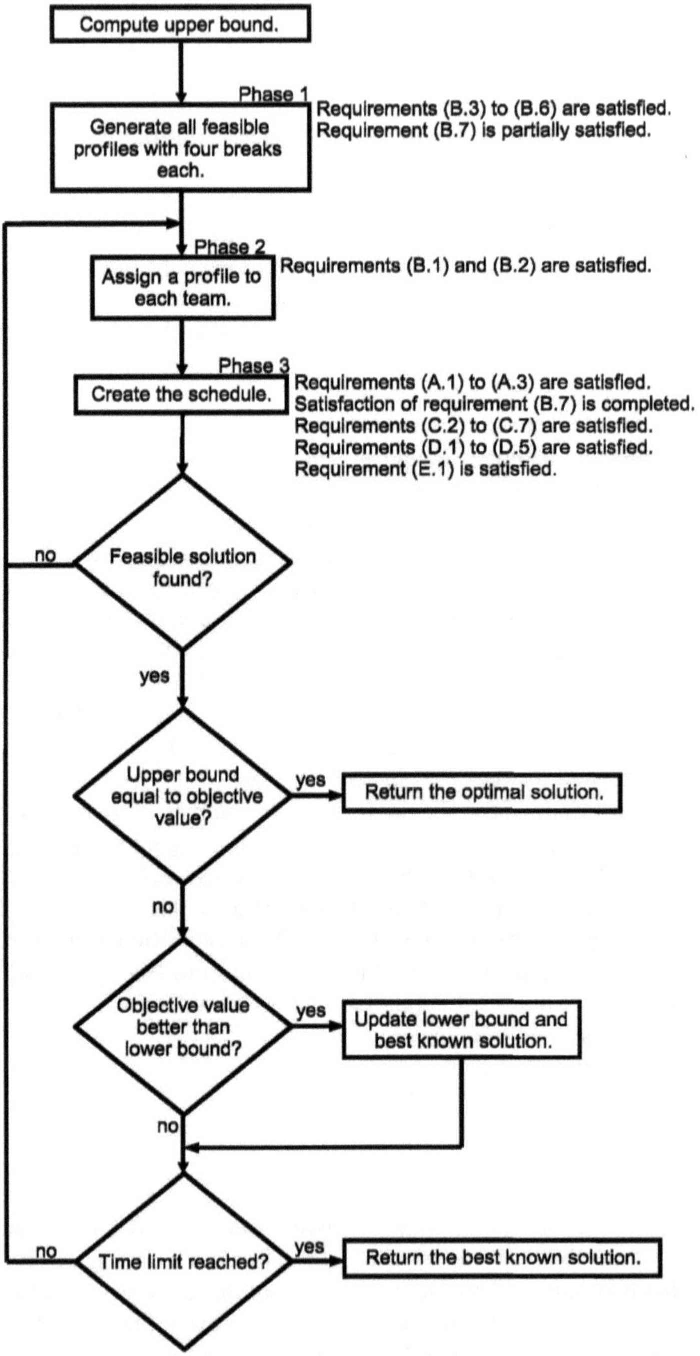

Fig. 5.5 Solution approach based on a three-phase decomposition scheme

that occurs in the first round of the second half of the tournament. Suppose there are no breaks in a profile. If a team plays at home in the first round, it will play at home (resp. away) in every odd (resp. even) round. After a break, this sequence is reversed: this same team will play at home in the following odd rounds and away in the following even rounds. Every additional break reverses this sequence in the same manner. Therefore, if a team playing at home in the first round has an odd number of breaks in the first half of the schedule, then it will play away in round 19 (i.e., in the last round of the first half). This team must play away in the first round of the second half because the schedule is mirrored; therefore, it will have a break in the first round of the second half.

Profiles satisfying constraints (B.4) have an even number of breaks because the first round is odd, and round 38 (the last) is even. Therefore, we consider only profiles with an even number of breaks in the first half. With constraints (B.5) and (B.6), this requirement implies that we must consider only profiles with the same even number of breaks and with as few breaks as possible. Since every team must have a different profile in every schedule, and there are only two profiles without breaks (i.e., those that perfectly alternate between home and away games), we must consider profiles with exactly two breaks in each tournament phase.

Profiles satisfying constraints (B.3) are those without breaks in the second, third, fourth, and last rounds. Because the schedule is mirrored, there is no break in the first phase's last round (round $2n-1$). Therefore, there are $2n-6$ rounds (those numbered from 5 to $2n-2$) in which the breaks may occur. To ensure that every team has one home break and one away break, as implied by requirement (B.5), the two rounds in which each team has a break must be both even or odd. There are $(2n-6)/2$ even rounds and $(2n-6)/2$ odd rounds in which the teams may have their two breaks. Therefore, there are

$$2 \cdot \binom{(2n-6)/2}{2} = \left(\frac{2n-6}{2} - 1\right) \frac{2n-6}{2}$$

possible break configurations. Moreover, there is exactly one profile starting with a home game for every possible break configuration, making the number of feasible profiles with exactly two breaks equal to $2 \cdot ((2n-6)/2 - 1)(2n-6)/2 = 84$ for $2n = 20$.

The profiles generated in this phase satisfy requirements (B.3)–(B.6). All these profiles have either nine home games and ten away games or ten home games and nine away games; therefore, the requirement (B.7) is fulfilled partially for such assignments. The complete satisfaction of this requirement is enforced by setting the game between teams of the same pair at the venue of the team playing ten home games in the first phase. The small number of feasible profiles with exactly two breaks each makes it possible to enumerate them entirely in this phase.

5.2.4.2 Phase 2: Profile Assignment

In this phase, two complementary profiles are randomly chosen and assigned to each pair of teams. Therefore, each pair of teams takes a different pair of profiles. Assigning complementary profiles to pairs of teams (instead of one profile to each team) makes it possible to satisfy requirements (B.1) and (B.2): the first profile in the pair is assigned to the odd indexed team starting at home. On the contrary, its complement is assigned to the even indexed team in the same pair.

5.2.4.3 Phase 3: Schedule Creation

At this point, a profile has been assigned to each team. In this phase, a simplified integer program derived from the formulation presented in Sect. 5.2.3 is built and solved by considering the profiles currently assigned to each team.

Because each team has been already assigned to a profile in the second phase, the variables $y_{e\ell}$ are no longer necessary, and constraints (5.29) and (5.30) no longer exist. Furthermore, the right-hand side of constraints (5.31)–(5.34) is set to either 0 or 1, depending on each team's profile. This characteristic greatly simplifies the formulation by trivially setting to zero one-half of all x_{ijk} variables. Consequently, this problem can be solved by a commercial solver in reasonable computation times.

This integer program enforces that requirements (A.1)–(A.3), (B.7), (C.1)–(C.7), (D.1)–(D.5), and (E.1) are satisfied. We first assume that it is feasible. Suppose its optimal value is equal to the upper bound on the number of classic games that can be played in double weekend rounds. In that case, the algorithm terminates with an optimal solution. If its optimal value is smaller than this upper bound but improves the best lower bound (i.e., the best known solution value), then the best solution found is updated. If the maximum time limit given to the algorithm is reached or another stopping criterion is met, then the best solution found is returned. Otherwise, if the integer program is infeasible or its optimal value is smaller than the upper bound but the time limit is not reached, the algorithm returns to the second phase to perform new profile assignments.

5.2.5 Development and Practical Experience

The complete optimization model and the software system coded in C++ were developed, tuned, and validated over 3 years. CBF and TV Globo staff participated actively in formulating the problem and validating its results. The optimization software could generate solutions in a few minutes of processing time on a standard processor using the mixed integer programming solver CPLEX 10.2. A database with historical tournament data supported the user in avoiding repetitions of schedules and situations observed in previous years, such as playing at home or away in the first round.

The optimization software was validated with data from the first division of the tournament's editions of 2005 and 2006, played by 22 and 20 teams. The schedules obtained by integer programming have been proven to be much better than those manually built and used in practice. The official schedules played in 2005 and 2006 violated some requirements, while the optimized schedules met all constraints. There were 156 breaks in the schedule played in 2005 and 172 breaks in 2006: both numbers were far greater than the minimum number of breaks established by the requirements imposed by CBF. At that time, the main objective was to find schedules in which a maximum number of games between teams from Rio de Janeiro and São Paulo could be broadcast by open TV channels. The new approach led to schedules in which all 56 of these most attractive games could be broadcast by open TV channels without conflicts. The ad hoc rules used for scheduling the 2005 and 2006 tournament editions made it possible to broadcast only 43 and 47 games, respectively.

The new system was first used in 2009 as the official scheduler for the Brazilian football tournament, with 20 teams in each division. Multiple schedules were provided to the users, who selected their preferred choice. New criteria have been proposed and introduced into the system during the decision process based on successive refinements of the solution as the decision-makers evaluated and filtered the different solutions. The organizers (CBF, sponsors, and security authorities) checked each scheduled proposed and imposed additional constraints (or removed existing constraints) to handle specific situations that might be desired to fine-tune the schedule. The final schedule was announced in January 2009. The tournament was the most attractive until that year, with four teams (Flamengo, Internacional, São Paulo, and Palmeiras) still in contention for the title when the last round started, with all games being played simultaneously. The title changed hands several times during almost 2 hours of play as the scores of the ten games underway changed. The goal that decided the tournament for Flamengo was scored only 20 minutes before the tournament's end. The championship was undecided until the last game ended, contrary to what had happened in previous years, when the winners were known up to four rounds before the end of the tournament, making the games of the last rounds completely uninteresting. Because of the fight for the title, many attractive games have been played in the last rounds, with high interest and large gate attendance. This scenario was partly due to the fair and balanced schedule of games, in which no team had specific advantages or disadvantages over the others.

The optimization system was used for the second time in 2010 (the tournament edition that was used to describe the problem formulation). Once again, the decision-makers were happy with the alternative schedules the system computed and the choices they could make. This edition was a challenging tournament to schedule. It had to be interrupted in June and July for the 2010 World Cup. Fewer dates were available for game playing. As a result, there were more midweek rounds and fewer weekend rounds, making it impossible to schedule all classic games in double weekend rounds. Therefore, the system was set to seek a schedule with a maximum number of classic games played in double weekend rounds. Furthermore, for the first time in many years, all G12 teams participated in the tournament, augmenting the number of constraints. The increase in the number of teams from the state of São Paulo also made it harder to find balanced schedules because these teams could benefit from short trips within the state. As for the previous year, the title was decided in the last round, with three teams (Fluminense, Corinthians, and Cruzeiro) still in contention for the title when their matches started. The goal that decided the tournament for Fluminense was scored 25 minutes before the end of the tournament.

The officers of CBF requested some extensions in the system in 2011 to enforce better practices and more transparency. In previous years, teams that were not on the course for the title in the last rounds of the tournament have been accused of facilitating their final games to avoid some of their local rivals winning the tournament. The best players of Grêmio did not play in the last game in 2009. Some people claimed that the reason was to facilitate a winning of Flamengo and avoid giving the championship to Internacional (Grêmio's historical local rival in the same state). A player of Palmeiras was criticized by its fans when he scored first against Fluminense in 2010 since this goal could open the path to Corinthians (Palmeiras' historical local rival in the same state) winning the title. New constraints were incorporated into the system to schedule all local classic games in the final rounds to avoid this behavior related to local rivalries that might affect the tournament outcome. The result was even better than expected, and the users were pleased: derbies played in the final rounds brought much emotion, and complaints of unfair play ceased. The decision-makers were pleased with the schedule produced by the system. Once again, three teams were still in contention for the title in the last round.

Football is a sport for which it is hard to predict results in advance, especially for long tournaments. Attempts to develop schedules that pair the strongest (i.e., the most likely title-contending) teams at the end of the tournament have failed because of the difficulty in accurately identifying them in advance. The approach developed for CBF that led to balanced schedules in which all teams face the same difficulties proved to be a good strategy to maximize fairness, balance, and the attractiveness of the games throughout the tournament.

5.3 Dynamic Scheduling of E-Sports Tournaments

This section presents two different and innovative aspects concerning other applications and models discussed in this book. First, we describe an application to electronic sports, not to more traditional sports. Second, the structure of the integer programming model used in the problem formulation is quite different from the others. We are not scheduling an entire tournament but just each of its rounds, one after the other, which makes the integer programming model small and easy to solve.

Professional traditional sports involve millions of fans and significant investments in players, broadcast rights, merchandising, and advertising, facing challenging logistics and optimization problems. On the other hand, amateur leagues involve fewer investments but require coordination and logistical efforts due to many tournaments and competitors. Electronic sports (e-sports) bring new challenges due to the different nature and designs of their tournaments.

As internet technologies and devices improve very quickly, electronic sports have had fast growth in market value and the number of participants. The General Administration of Sport of China officially launched e-sports as a formal sport in 2003. E-sports started as an exhibition sport at the Asian Games in 2018. The Olympic Council of Asia announced that e-sports would become one of the 37 official sports in the 2022 Hangzhou Asian Games (delayed to 2023).

We have seen in this book that good schedules play a vital role in sports tournaments. They can significantly improve their fairness and logistics while reducing the organizers' costs. Scheduling methods initially developed for traditional sports are only sometimes appropriate for e-sports due to their competitions' nature and the tournaments' formats.

E-sports differ from traditional sports in several aspects. First, there are differences in the athletes' physicality: e-sports involve movements using small muscles, whereas traditional sports require large muscles. Second, sport is conceived as an area organized and regulated by institutions (such as FIFA in the case of football). Although one of the five criteria defined by the Global Association of International Sports Federations states that "sport should not rely on equipment that a single supplier provides," e-sports rely on commercial products owned and governed by private organizations. Third, traditional sports and e-sports competitions are organized in distinct tournament formats in different environments: while traditional sports competitions are played usually in open fields, courts, or large arenas, most major e-sports tournaments are held in venues that can hold thousands of spectators or are streamed online.

We have seen in Sect. 1.1.3.3 that the Swiss system is an appropriate tournament design for e-sports. This section considers a Swiss-system tournament with $2n$ teams organized in a fixed number R of rounds. Every team plays exactly one game in each round. The games in each round are played in n sequential, non-overlapping time slots. The best-ranked team after the last round is the winner.

In the following, we describe and evaluate a dynamic approach for scheduling Swiss-system e-sports tournaments with this structure. It uses optimization methods to pair the teams that will meet in each round and determine the time slots in which the games will be played. We first describe Colley's method for rating teams with opponents of different strengths (i.e., teams that have played against opponents with different strengths in the previous rounds). Next, we present the dynamic approach developed for team pairing and game scheduling in each tournament round. The ratings computed by Colley's method are used to formulate the associated integer programming problem. Numerical results validate the approach.

5.3.1 Rating Teams with Opponents of Different Strengths

We observed in Sect. 1.1.3.3 that the Swiss system seeks to create pairings of competitors with a similar score (or rating) in each round.

Although simple statistics such as the number of wins would suffice to produce a fair ranking if all competitors in a tournament played opponents with the same (or similar) strength, such as in a round-robin tournament, this is different for the Swiss system. First, each competitor plays against a small subset of competitors (and not against all contenders). Second, the competitors play against opponents of different strengths. Consequently, establishing a ranking is more complex if consistency is an objective.

Remark 5.1 We illustrate how simple statistics may lead to similar rankings for competitors that play against opponents with different strengths. For example, two competitors, A and B, may have three wins each over very different competitors. Suppose that competitor A defeated three strong competitors, while competitor B defeated three weak competitors. If only wins were counted, then both would have the same ranking (i.e., three wins), although competitor A seems stronger than B. ■

Therefore, Colley's ranking method was used to overcome the difficulties of ranking the strength of all competitors in the tournament.

The approach is simple and considers only wins and defeats. It avoids the reliance on scores that generate some dependence on margins that may be trickier, mainly when the competitors have opponents with divergent strengths, as is the case with the Swiss system. Using only wins and defeats eliminates any bias toward history or tradition, eliminates the need to invoke some ad hoc means of deflating runaway scores, and eliminates any other ad hoc adjustments, such as home/away adjustments for extramural tournaments.

The rating r_i^k of each competitor $i = 1, \ldots, 2n$ at some round $k = 1, \ldots, R$ of the competition is given by

$$r_i^k = \frac{1 + wins_i^k}{2 + total_i^k}, \tag{5.38}$$

where $wins_i^k$ and $total_i^k$ are, respectively, the number of wins and the total number of games played by competitor i until round k. Therefore, at the beginning of the season, all competitors have an equal rating of 1/2. After winning one game in the first round, a competitor improves its rating to 2/3, while the rating of a losing competitor becomes 1/3.

The number of wins of competitor i in Eq. (5.38) may be rewritten as

$$wins_i^k = \frac{wins_i^k - defeats_i^k}{2} + \frac{total_i^k}{2} = \frac{wins_i^k - defeats_i^k}{2} + 0.5 \cdot total_i^k, \tag{5.39}$$

where $defeats_i^k$ denotes the number of defeats of competitor i until round k and the last term $0.5 \cdot total_i^k$ corresponds to the sum of the ratings of $total_i^k$ random competitors. If these random competitors are replaced by the first $total_i^k$ opponents of competitor i, and we use their actual ratings, then we obtain the effective number of wins of competitor i:

$$effwins_i^k = \frac{wins_i^k - defeats_i^k}{2} + \sum_{j=1}^{total_i^k} r_{opp(i,j)}^j, \qquad (5.40)$$

where $opp(i, j)$ is the opponent of competitor i in round j. The second term (the summation) in the right-hand side of the Eq. (5.40) is the adjustment for the strength of the specific schedule played by competitor i.

All competitors' ratings are initialized with 1/2 and computed iteratively at each round up to a specific numerical tolerance using Eq. (5.40).

A more elegant and efficient solution approach consists in rewriting Eqs. (5.38) and (5.40) as follows, by setting $wins_i^k = effwins_i^k$ for every competitor i, as in the outcome of the iterative approach:

$$(2 + total_i^k) r_i^k - \sum_{j=1}^{total_i^k} r_{opp(i,j)}^j = 1 + \frac{wins_i^k - defeats_i^k}{2}, \qquad i = 1, \ldots, 2n, \qquad (5.41)$$

which is, for each round $k = 1, \ldots, R$, a system of $2n$ linear equations with $2n$ variables, each associated with one competitor. These equations may be written in matrix form as follows:

$$C^k \cdot r^k = b^k, \qquad (5.42)$$

where r^k is the $2n \times 1$ column vector formed by all ratings $r_i^k, i = 1, \ldots, 2n$, and b^k is the $2n \times 1$ column vector associated with the right-hand side of Eq. (5.41):

$$b_i^k = 1 + \frac{wins_i^k - defeats_i^k}{2}, \qquad i = 1, \ldots, 2n. \qquad (5.43)$$

$C^k = \{c_{ij}^k\}$ is the $2n \times 2n$ Colley matrix with coefficients

$$c_{ii}^k = 2 + total_i^k, \qquad i = 1, \ldots, 2n, \qquad (5.44)$$

$$c_{ij}^k = -games_{ji}^k, \qquad i, j = 1, \ldots, 2n : i \neq j, \qquad (5.45)$$

where $games^k_{ji} = games^k_{ij}$ is the number of times competitor i played with competitor j until round k.

Solving the linear system (5.42) with the coefficients given by Eqs. (5.43)–(5.45) is the method for computing the ratings r_i^k for every competitor $i = 1, \ldots, 2n$ in each round k. Existing solvers can directly solve it by Cholesky decomposition and back-substitution because each matrix C^k is symmetric, real-valued, and positive definite. The Colley matrix method conserves an average rating of 1/2. All ratings belong to the interval (0, 1).

In addition to other characteristics mentioned above, some advantages of this approach are that it is reproducible, uses a minimum of assumptions, ignores runaway scores, and, very importantly, adjusts for the strength of the schedule.

Remark 5.2 Colley's ranking method assumes that each game has a winner. Another ranking method should be used for pairing the competitors in the case of sports (such as football) tournaments whose games may end up in a draw. ■

5.3.2 Dynamic Scheduling: Team Pairing and Game Scheduling

The attractiveness (demand or interest) of a game in a league depends on the competitive balance of the tournament, measured by the uncertainty of its outcome: fans enjoy sports events more if the winner is challenging to predict. In other words, the more similar the competitors' winning chances are, the more exciting the tournament is.

The ratings produced by Colley's method described in the previous section will be used for pairing dynamically round-by-round the teams participating in the tournament following the Swiss-system design. To maximize the tournament's attractiveness, pairings in each round will involve teams with similar ratings as much as possible. To assess the tournament's competitive balance, we define the *unattractiveness* of the game between teams i and j in round k by the square of the difference between their ratings. The more similar the ratings of the two teams are, the more attractive (or less unattractive) the game between them is:

$$u^k_{ij} = (r_i^k - r_j^k)^2. \tag{5.46}$$

The more balanced the game between teams i and j is, the closer the value of u^k_{ij} is to zero. Therefore, we seek to minimize the sum of u^k_{ij} over the games played in each round $k = 1, \ldots, R$ to increase the tournament's competitive balance.

There are exactly n games in each round, and they should be played consecutively in exactly n non-overlapping time slots. In other words, there is no overlapping between any pair of games played in the same round. Every game can be followed live by the attendance or on the internet. For fairness, to give the approximately same resting time for all teams between two consecutive rounds, bounds are imposed between the positions of the time slots in which any team's games are scheduled in any two consecutive rounds. An external parameter, D, set by the organizers establishes the maximum difference (or distance) between the time slots of the games played by any team in rounds k and $k-1$. Since the only feasible solution for $D = 0$ would correspond to the repetition in round k of the same games played in round $k-1$ in the same order, this case may be disregarded. $D = 1$ corresponds to the case where the time slot of the game played by each team may change by at most one unit between rounds k and $k-1$. $D \geq n-1$ corresponds to the unconstrained case.

In addition, to maximize interest in the tournament, repetitions of the same game will be avoided except when necessary to ensure feasibility. Another external parameter, L, set by the organizers, defines the minimum number of rounds between two game occurrences involving the same pair of teams. $L = 0$ corresponds to an unconstrained problem (i.e., repetitions arbitrarily allowed). $L = R-1$ corresponds to the more challenging case in terms of feasibility, where there is at most one game between any pair of teams (i.e., no game repetitions are allowed).

We consider a formulation that simultaneously optimizes team pairing (i.e., opponents selection) and game scheduling (i.e., the order in which the games are played in each round). We define the following decision variable:

$$x_{ijt}^{k} = \begin{cases} 1, & \text{if team } i \text{ plays with team } j \text{ in time slot } t \text{ of round } k, \\ 0, & \text{otherwise.} \end{cases} \tag{5.47}$$

The goal consists in maximizing the interest (which amounts to minimizing the unattractiveness) of the games in each round $k = 1, \ldots, R$, which can be achieved by scheduling attractive games between teams with similar ratings. The number R of rounds is another external parameter set by the organizers and is known beforehand. In the first round, the pairings can be made randomly. The problem to be solved in each of the following rounds $k = 2, \ldots, R$ may be formulated as follows:

$$\min \sum_{i=1}^{2n} \sum_{\substack{j=1 \\ j \neq i}}^{2n} \sum_{t=1}^{n} u_{ij}^{k} x_{ijt}^{k} \tag{5.48}$$

subject to:

$$x_{ijt}^{k} = x_{jit}^{k}, \qquad i, j = 1, \ldots, 2n : i \neq j; \quad t = 1, \ldots, n, \tag{5.49}$$

$$\sum_{\substack{i=1 \\ i \neq j}}^{2n} \sum_{t=1}^{n} x_{ijt}^{k} = 1, \qquad j = 1, \ldots, 2n, \tag{5.50}$$

$$\sum_{i=1}^{2n} \sum_{\substack{j=1 \\ j \neq i}}^{2n} x_{ijt}^k = 2, \qquad t = 1, \ldots, n, \tag{5.51}$$

$$\sum_{t=1}^{n} x_{ijt}^k \leq 1 - \sum_{\ell=1}^{\min\{L,k-1\}} \sum_{t=1}^{n} \bar{x}_{ijt}^{k-\ell}, \qquad i, j = 1, \ldots, 2n : i \neq j, \tag{5.52}$$

$$\sum_{\substack{j=1 \\ j \neq i}}^{2n} \sum_{t=1}^{n} t(x_{ijt}^k - \bar{x}_{ijt}^{k-1}) \leq D, \qquad i = 1, \ldots, 2n, \tag{5.53}$$

$$\sum_{\substack{j=1 \\ j \neq i}}^{2n} \sum_{t=1}^{n} t(\bar{x}_{ijt}^{k-1} - x_{ijt}^k) \leq D, \qquad i = 1, \ldots, 2n, \tag{5.54}$$

$$x_{ijt}^k \in \{0, 1\}, \qquad i, j = 1, \ldots, 2n; i \neq j; \quad t = 1, \ldots, n. \tag{5.55}$$

In (5.48), we minimize the objective function, which is the sum of the squared differences between the ratings of the teams paired in round k. The symmetry constraints (5.49) establish that the game between teams i and j is the same as the one between teams j and i. Constraints (5.50) ensure that every team plays precisely once in round k. The no-overlap constraint (5.51) guarantee that there will be exactly one game scheduled in each of the n time slots of round k. Inequalities (5.52) are the no-repeaters constraints: they ensure that there should be at least L rounds between any two games between teams i and j ($L = 0$ means that two games between teams i and j may take place in two consecutive rounds). Constraints (5.53)–(5.54) impose that the distance between the positions of the games played by each team in the previous and current rounds is bounded from above by a given parameter D (the maximum distance). Constraints (5.55) guarantee the integrality of the decision variables.

Remark 5.3 We recall that model (5.48)–(5.55) above denotes the problem solved at each round $k = 2, \ldots, R$, and that the index k is fixed in the model. Note that each $\bar{x}_{ijt}^{k-\ell}$ in this formulation does not represent a problem variable, but instead, the actual value of the variable $x_{ijt}^{k-\ell}$, already computed for one of the previous rounds $k - \ell$. ■

5.3.3 Numerical Results

All algorithms described in this section were implemented in C++ and integrated with CPLEX.

Two sets of computational experiments are reported. Section 5.3.3.1 reports results for real-life data collected from the 2020 Honor of Kings World Champion Cup (KCC2020). Section 5.3.3.2 reports results for randomly generated problems of larger size.

5.3.3.1 Simulated Results for Dynamic Scheduling from Realistic Data

The dynamic scheduling approach based on the Swiss system discussed in Sect. 5.3.2 was applied to real-life data collected from the 2020 Honor of Kings World Champion Cup (KCC2020) with 12 teams. This tournament was organized in two groups, A and B, of six teams each. Each group played a single round-robin tournament. The actual results for the games of groups A and B of KCC2020 appear in Tables 5.5 and 5.6, respectively.

The dynamic scheduling approach was applied to groups A and B. All teams had the same ratings in the first round. Therefore, the first pairings were randomly arranged (the initial ratings could also have been taken from the previous tournament in which they participated together). The actual results in Tables 5.5 and 5.6 have been used to simulate the result of each game with the dynamic scheduling approach.

Figures 5.6 and 5.7 illustrate how the ratings and rankings of the six teams in groups A and B, respectively, evolve along ten rounds scheduled by the dynamic approach, corresponding to twice the number of rounds in a single round-robin tournament with six teams. Each cell displays the rating of the corresponding team and round (number of points for the single round-robin column). For any round, cells with the same rating correspond to teams tied in the same position or with the same ranking. The results for group B are very stable: round 3 has

Table 5.5 Actual results of the games of the six teams in group A of KCC2020

	QG Happy	AG Super Play	eStar Pro	Team WE	Rox Gaming
Turnso Gaming	2×1	2×1	1×2	2×0	2×0
QG Happy		2×0	2×0	0×2	2×0
AG Super Play			2×0	2×0	2×0
eStar Pro				2×0	2×0
Team WE					2×0

Table 5.6 Actual results of the games of the six teams in group B of KCC2020

	ViCi	Douyu	TTG X-Quest	Ghost Owl	Esports of Macau
Mighty Tiger	2×1	1×2	2×0	2×1	2×0
ViCi		2×1	2×0	1×2	2×0
Douyu			1×2	2×0	2×0
TTG X-Quest				2×0	2×0
Ghost Owl					2×0

Rank	Round 1	Round 2	Round 3	Round 4	Round 5	Round 6	Round 7	Round 8	Round 9	Round 10 (final)	Round robin (result)
1st	Turnso Gaming **0.625**	QG Happy **0.766**	QG Happy **0.653**	Turnso Gaming **0.687**	Turnso Gaming **0.687**	Turnso Gaming **0.712**	QG Happy **0.661**	Turnso Gaming **0.691**	Turnso Gaming **0.720**	Turnso Gaming **0.714**	Turnso Gaming **4 wins**
2nd	QG Happy **0.625**	eStar Pro **0.566**	Turnso Gaming **0.625**	QG Happy **0.666**	QG Happy **0.562**	QG Happy **0.612**	Turnso Gaming **0.651**	QG Happy **0.621**	QG Happy **0.625**	QG Happy **0.571**	QG Happy **3 wins**
3rd	AG Super Play **0.625**	Turnso Gaming **0.500**	eStar Pro **0.596**	eStar Pro **0.645**	eStar Pro **0.562**	AG Super Play **0.575**	eStar Pro **0.581**	eStar Pro **0.586**	eStar Pro **0.529**	eStar Pro **0.571**	eStar Pro **3 wins**
4th	eStar Pro **0.375**	AG Super Play **0.500**	AG Super Play **0.546**	AG Super Play **0.479**	AG Super Play **0.562**	eStar Pro **0.512**	AG Super Play **0.530**	AG Super Play **0.560**	AG Super Play **0.613**	AG Super Play **0.571**	AG Super Play **3 wins**
5th	Team WE **0.375**	Team WE **0.433**	Team WE **0.375**	Team WE **0.333**	Team WE **0.437**	Team WE **0.412**	Team WE **0.429**	Team WE **0.399**	Team WE **0.375**	Team WE **0.428**	Team WE **2 wins**
6th	Rox Gaming **0.375**	Rox Gaming **0.233**	Rox Gaming **0.203**	Rox Gaming **0.187**	Rox Gaming **0.187**	Rox Gaming **0.175**	Rox Gaming **0.146**	Rox Gaming **0.140**	Rox Gaming **0.136**	Rox Gaming **0.142**	Rox Gaming **0 wins**

Fig. 5.6 Group A of KCC2020

Rank	Round 1	Round 2	Round 3	Round 4	Round 5	Round 6	Round 7	Round 8	Round 9	Round 10 (final)	Round robin (result)
1st	Mighty Tiger Gaming 0.625	Mighty Tiger Gaming 0.711	Mighty Tiger Gaming 0.796	Mighty Tiger Gaming 0.687	Mighty Tiger Gaming 0.687	Mighty Tiger Gaming 0.712	Mighty Tiger Gaming 0.742	Mighty Tiger Gaming 0.775	Mighty Tiger Gaming 0.720	Mighty Tiger Gaming 0.714	Mighty Tiger Gaming 4 wins
2nd	ViCi Gaming 0.625	ViCi Gaming 0.711	ViCi Gaming 0.653	ViCi Gaming 0.666	ViCi Gaming 0.562	ViCi Gaming 0.612	ViCi Gaming 0.651	ViCi Gaming 0.621	ViCi Gaming 0.625	ViCi Gaming 0.571	ViCi Gaming 3 wins
3rd	Douyu Gaming 0.625	Douyu Gaming 0.488	Douyu Gaming 0.546	Douyu Gaming 0.645	Douyu Gaming 0.562	Douyu Gaming 0.575	Douyu Gaming 0.529	Douyu Gaming 0.560	Douyu Gaming 0.613	Douyu Gaming 0.571	Douyu Gaming 3 wins
4th	TTG X-Quest 0.375	Ghost Owl Gaming 0.488	TTG X-Quest 0.425	TTG X-Quest 0.479	TTG X-Quest 0.562	TTG X-Quest 0.512	TTG X-Quest 0.489	TTG X-Quest 0.502	TTG X-Quest 0.529	TTG X-Quest 0.571	TTG X-Quest 3 wins
5th	Ghost Owl Gaming 0.375	TTG X-Quest 0.355	Ghost Owl Gaming 0.403	Ghost Owl Gaming 0.333	Ghost Owl Gaming 0.437	Ghost Owl Gaming 0.412	Ghost Owl Gaming 0.438	Ghost Owl Gaming 0.406	Ghost Owl Gaming 0.375	Ghost Owl Gaming 0.428	Ghost Owl Gaming 2 wins
6th	Esports of Macau China 0.375	Esports of Macau China 0.244	Esports of Macau China 0.175	Esports of Macau China 0.187	Esports of Macau China 0.187	Esports of Macau China 0.175	Esports of Macau China 0.146	Esports of Macau China 0.133	Esports of Macau China 0.136	Esports of Macau China 0.142	Esports of Macau China 0 wins

Fig. 5.7 Group B of KCC2020

Table 5.7 Games scheduled in the five first rounds by the dynamic approach for group A of KCC2020

	Rounds				
Teams	1	2	3	4	5
Turnso Gaming	Team WE	eStar Pro	QG Happy	AG Super Play	Rox Gaming
QG Happy	eStar Pro	AG Super Play	Turnso Gaming	Rox Gaming	Team WE
AG Super Play	Rox Gaming	QG Happy	Team WE	Turnso Gaming	eStar Pro
eStar Pro	QG Happy	Turnso Gaming	Rox Gaming	Team WE	AG Super Play
Team WE	Turnso Gaming	Rox Gaming	AG Super Play	eStar Pro	QG Happy
Rox Gaming	AG Super Play	Team WE	eStar Pro	QG Happy	Turnso Gaming

Table 5.8 Games scheduled in the five first rounds by the dynamic approach for group B of KCC2020

	Rounds				
Teams	1	2	3	4	5
Mighty Tiger	Ghost Owl	TTG X-Quest	ViCi	Douyu	Esports Macau
ViCi	TTG X-Quest	Douyu	Mighty Tiger	Esports Macau	Ghost Owl
Douyu	Esports Macau	ViCi	Ghost Owl	Mighty Tiger	TTG X-Quest
TTG X-Quest	ViCi	Mighty Tiger	Esports Macau	Ghost Owl	Douyu
Ghost Owl	Mighty Tiger	Esports Macau	Douyu	TTG X-Quest	ViCi
Esports of Macau	Douyu	Ghost Owl	TTG X-Quest	ViCi	Mighty Tiger

already obtained without ties the same ranking found in round 10, which is the same obtained by the round-robin scheduling. For group A, the situation is slightly different. While the fifth and sixth places have been stable since the first round, there are alternations in the first four positions. The fourth (without ties) and fifth rounds have already obtained the same ranking found in round 10 (which once again is the same obtained by the initial round-robin scheduling). However, some oscillations were still observed in rounds 6 and 7.

These results confirm that the outcome of tournaments played with the Swiss-system design is sensitive to the number of rounds played and that the latter becomes a critical parameter to be set by the organizers.

The games scheduled in the five first rounds by the dynamic approach are displayed in Tables 5.7 and 5.8.

5.3.3.2 Results for Randomly Generated Larger Tournaments

This section reports the application of the dynamic scheduling approach to larger tournaments with up to $2n = 80$ teams. We evaluate its computational efficiency

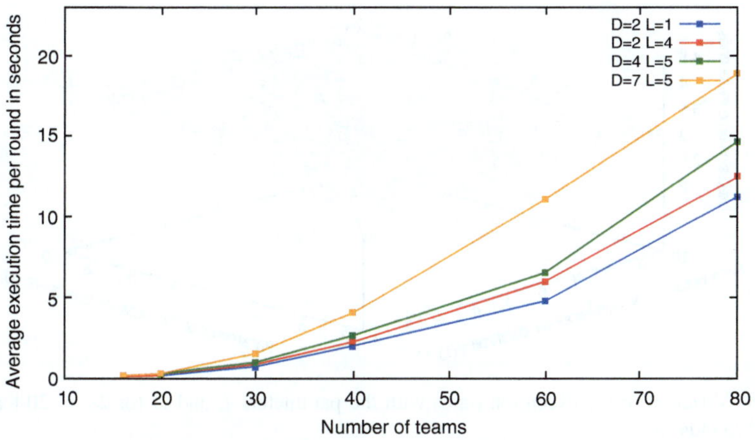

Fig. 5.8 Average execution times per round with the increase in the number of teams

and assess the numerical results' consistency. Two parameters are considered in the numerical analysis:

- L is the minimum number of rounds between two game occurrences involving the same pair of teams. $L = 0$ corresponds to the unconstrained case, while for $L \geq R - 1$ no repetitions are allowed.
- D is the maximum distance between the positions of the games played by any team in any pair of consecutive rounds. $D \geq n - 1$ corresponds to the unconstrained case, while for $D = 1$, the position of the game played by each team may change by at most one unit between any two consecutive rounds.

Simulations have been performed for $2n = 16, 20, 30, 40, 60, 80, 100$, and the following values of the above parameters: $L = 1, \ldots, 10$, and $D = 1, \ldots, 15$. Maximum winning probabilities have been used to randomly determine each game's winner along the tournament simulation.

Figure 5.8 displays the average execution times per round of the dynamic scheduling approach running for $R = n$ rounds as the number of teams ranges from $2n = 16$ to $2n = 80$. We observe that the approach is feasible in practice. The execution times are small, although they increase for larger values of L and D.

The influence of parameters L and D in the execution times is further illustrated in Fig. 5.9. The execution times become more significant as the distance parameter D increases, while there is less variation with the repetition parameter L.

These two parameters, related to fairness and attractiveness, give more flexibility to the tournament organizers. In practice, large values of L (to avoid repetitions of the same game) and $D \approx 2$ are adequate. The integer programming problem in each round can be solved fast, independently of the parameter values set by the tournament organizer.

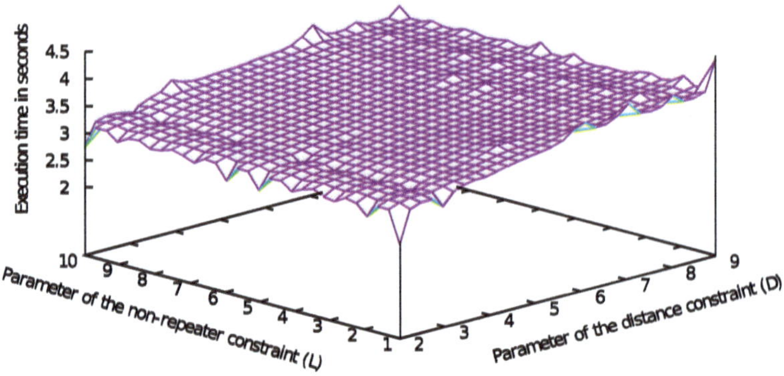

Fig. 5.9 Variation of the execution times with the parameters L and D for $2n = 20$ teams and $R = 10$ rounds

Figure 5.10a and b illustrate for two tournaments with $2n = 16$ and $2n = 20$ teams, respectively, how the teams ranked in the first six positions evolve from round $\lceil \log_2 2n \rceil$ to round $2n - 1$. They show that the teams ranked in the first and second positions do not change along all these rounds (i.e., after the first $\lceil \log_2 2n \rceil$ rounds), although some changes occur between the third and sixth positions. In other words, convergence and stability can be obtained in $\lceil \log_2 2n \rceil$ or a few more rounds, where $2n$ is the number of competitors. The rating, pairing, and scheduling methods seem appropriate to lead to the tournament's winner.

5.3.4 *Concluding Remarks*

Electronic sports have had fast growth in market value and the number of participants in recent years. At the same time, e-sports generate more revenues and become official sports in events like the Asian Games. Tournaments of e-sports attract progressively more attention and pay higher total prize pools.

This section reported an application of optimization methods to the scheduling of e-sports tournaments, introducing a dynamic scheduling strategy for e-sports tournaments based on the Swiss-system design. The two parameters used in the optimization model are related to fairness and attractiveness, giving more flexibility to the tournament organizers.

The integer programming problem in each round can be solved fast, independently of the parameter values set by the tournament organizer. The numerical results show that convergence and stability can be obtained in $\lceil \log_2 2n \rceil$ or a few more rounds, where $2n$ is the number of competitors.

(a)

Rank	Round 4	Round 5	Round 6	Round 7	Round 8	Round 9	Round 10	Round 11	Round 12	Round 13	Round 14	Round 15
1st	Team 16 0.930	Team 16 0.980	Team 16 0.989	Team 16 0.933	Team 16 0.948	Team 16 0.938	Team 16 0.863	Team 16 0.885	Team 16 0.909	Team 16 0.932	Team 16 0.932	Team 16 0.889
2nd	Team 1 0.759	Team 1 0.817	Team 1 0.820	Team 1 0.866	Team 1 0.891	Team 1 0.908	Team 1 0.847	Team 1 0.811	Team 1 0.780	Team 1 0.795	Team 1 0.823	Team 1 0.777
3rd	Team 15 0.670	Team 2 0.686	Team 14 0.708	Team 14 0.760	Team 14 0.729	Team 14 0.691	Team 14 0.711	Team 13 0.738	Team 14 0.765	Team 14 0.746	Team 14 0.714	Team 14 0.729
4th	Team 5 0.662	Team 14 0.627	Team 2 0.629	Team 15 0.663	Team 15 0.626	Team 4 0.635	Team 4 0.710	Team 14 0.732	Team 13 0.700	Team 13 0.657	Team 13 0.682	Team 13 0.720
5th	Team 12 0.608	Team 15 0.618	Team 15 0.609	Team 2 0.575	Team 13 0.593	Team 13 0.631	Team 13 0.690	Team 4 0.688	Team 4 0.648	Team 4 0.616	Team 7 0.593	Team 4 0.606
6th	Team 2 0.590	Team 5 0.614	Team 12 0.576	Team 3 0.559	Team 4 0.592	Team 15 0.574	Team 15 0.535	Team 2 0.545	Team 3 0.548	Team 3 0.585	Team 4 0.588	Team 3 0.600

(b)

Rank	Round 5	Round 6	Round 7	Round 8	Round 9	Round 10	Round 11	Round 12	Round 13	Round 14	Round 15	Round 16	Round 17	Round 18	Round 19
1st	Team 14 0.945	Team 14 0.864	Team 14 0.904	Team 14 0.934	Team 14 0.952	Team 14 0.942	Team 14 0.909	Team 14 0.928	Team 14 0.944	Team 14 0.901	Team 14 0.856	Team 14 0.88	Team 14 0.901	Team 14 0.922	Team 14 0.933
2nd	Team 16 0.751	Team 16 0.823	Team 16 0.860	Team 16 0.803	Team 16 0.840	Team 16 0.848	Team 16 0.894	Team 16 0.911	Team 16 0.927	Team 16 0.868	Team 16 0.834	Team 16 0.812	Team 16 0.836	Team 16 0.857	Team 16 0.868
3rd	Team 5 0.737	Team 5 0.811	Team 19 0.778	Team 5 0.801	Team 5 0.754	Team 5 0.689	Team 1 0.706	Team 19 0.729	Team 19 0.752	Team 19 0.777	Team 19 0.797	Team 19 0.771	Team 19 0.795	Team 19 0.819	Team 19 0.806
4th	Team 12 0.721	Team 19 0.746	Team 5 0.749	Team 19 0.744	Team 19 0.722	Team 19 0.671	Team 19 0.703	Team 1 0.678	Team 1 0.654	Team 5 0.657	Team 5 0.675	Team 5 0.707	Team 5 0.702	Team 5 0.727	Team 5 0.741
5th	Team 19 0.710	Team 7 0.681	Team 7 0.630	Team 12 0.665	Team 1 0.681	Team 1 0.658	Team 10 0.681	Team 5 0.645	Team 10 0.624	Team 10 0.652	Team 13 0.666	Team 17 0.676	Team 17 0.655	Team 17 0.643	Team 1 0.620
6th	Team 7 0.634	Team 12 0.661	Team 12 0.603	Team 1 0.625	Team 12 0.635	Team 10 0.650	Team 5 0.666	Team 10 0.644	Team 5 0.616	Team 13 0.632	Team 17 0.645	Team 13 0.642	Team 13 0.625	Team 13 0.610	Team 17 0.617

Fig. 5.10 Teams ranked in the first six positions from round $\lceil log_2 2n \rceil$ to round $2n - 1$ for $D = 2$ and $L = 4$. (a) $n = 16$. (b) $n = 20$

5.4 Related Literature

The results reported in Sect. 5.1 for the minimum weighted carry-over effects problem appeared in Guedes and Ribeiro (2011).

The problem of minimizing the carry-over effects was proposed originally by Russell (1980), where a construction algorithm was given for the (unweighted) minimum carry-over effects problem that generates schedules matching its lower bound when $2n$ is a power of two. The method proposed by Anderson (1999) obtained solutions that were still the best known for the unweighted instances at the time of the original publication of the content of this chapter, making use of algebraic structures called starters to generate schedules; see Dinitz (1996). The approach presumed that a suitable starter is known beforehand, which may imply huge computation times.

Some authors attribute minor consequences to carry-overs in collective sports (Goossens and Spieksma, 2009a). However, Flatberg et al. (2009) discussed the real-life case of a football league in Norway, in which carry-overs determined by one specific team and player strongly affected the outcome of the competition. In addition, Goodbread (2010) reported the case of US college football where a team was repeatedly scheduled against teams that had byes the week before (i.e., idle-rested opponents), consequently with a very unattractive schedule.

Trick (2000) developed a constraint programming method that made it possible to prove the optimality of Russell's method for $2n = 6$. Henz et al. (2004) improved the solution obtained by the previous approach for $2n = 12$, also using constraint programming. Miyashiro and Matsui (2006) developed a time-consuming heuristic based on random permutations of the rounds of schedules created by the circle method of Kirkman (1847). Their procedure obtained the same results as the two previously cited algorithms for $2n \leq 12$ and a better value for $2n = 14$. By adding constraints to the formulation proposed in Trick (2000), they also improved their best results for $2n = 10$ and $2n = 12$ and found the best known results for these two last values of $2n$ at the time of the publication of their results. They reported more than 2 days of computation times for $2n \geq 18$. However, their heuristic does not explore different strategies to generate the initial schedules, which limits the quality of the solutions it can provide.

The circle method generates schedules with maximum carry-over effect; see Lambrechts et al. (2018).

The heuristic proposed in Sect. 5.1 for the minimum weighted carry-over effects value problem was based on the hybridization of a multistart strategy with the iterated local search (ILS) metaheuristic; see Sect. 3.4.5. The first approach for obtaining starting schedules for the construction phase is the circle (or polygon) method, which gives the so-called canonical 1-factorization; see also Sect. 2.2. The second approach can be applied whenever $2n$ is a multiple of four; see de Werra (1980) and de Werra (1981). The rules used for selecting and placing rounds of the starting schedule in the incomplete solution under construction mimic those in the well known nearest neighbor heuristic and arbitrary insertion heuristics for the traveling salesman problem; see Lawler et al. (1985) and Gutin and Punnen (2002).

The original (unweighted) carry-over effects minimization problem does not consider any information concerning the relative strengths of the teams: all carry-over effects have the same unit weight. However, in real sports competitions, the organizers likely have an initial estimate of how well a specific team should perform in the competition, considering, e.g., its ranking in last year's competition, its tradition, the quality of its players, and the support of its fans in home games. Noronha et al. (2007) provided a concrete example of a competition (the Chilean national football tournament) in which the teams are classified a priori according to some of these criteria to build a fair schedule.

The local search used by Algorithm 16 followed the variable neighborhood descent strategy of Mladenovic and Hansen (1997); see also Hansen and Mladenović (2002). The best improving move in each neighborhood is applied to the current solution. The neighborhood structures mentioned in Sect. 5.1.3.2 have been described originally in Ribeiro and Urrutia (2007a) and Costa et al. (2008, 2012); see Sect. 3.5 for their detailed description. They were later unified by Januario and Urrutia (2016) and Januario et al. (2016). The ejection chain mechanism was proposed originally by Glover (1996). The ILS parameter is reset whenever the current solution is updated, following the same strategy already used in Ribeiro and Urrutia (2007a). The perturbation strategy based on using destructive neighborhoods to generate new solutions was proposed by Glover and Laguna (1997) and by Glover et al. (2000).

The quality of the results reported for the weighted minimum carry-over effects value problem confirmed previous successful cases of the hybridization of the ILS metaheuristic with multistart strategies, in particular in the context of scheduling problems in sports; see e.g. Duarte et al. (2007b) and Ribeiro and Urrutia (2007a).

The work reported in Sect. 5.2 appeared in Ribeiro and Urrutia (2012). A preliminary version of this work appeared in Ribeiro and Urrutia (2007b), where a bicriteria integer programming approach for scheduling the annual tournament of the first division was presented. There, a solution minimizing the number of breaks and maximizing the number of games that open TV channels could broadcast was sought.

Nurmi et al. (2010) observed that, despite the relevance of good schedules for the success and attractiveness of a tournament, only a few professional leagues had adopted optimization models and software. This situation is likely due both to the hardness of the problem and to some fuzzy preference restrictions and criteria that can hardly be described, but also to the resistance of teams and leagues to using new tools that introduce modern techniques in sports management.

Before this work, Bartsch et al. (2006) applied heuristics and branch-and-bound techniques to schedule the professional football leagues of Austria and Germany. Della Croce and Oliveri (2006) adapted the integer programming decomposition approach of Nemhauser and Trick (1998) for scheduling the Italian football league. Durán et al. (2007, 2009) used integer programming to schedule the Chilean football league. Improvements to their solution approach have been proposed by Noronha et al. (2007), see also Durán et al. (2012). Goossens and Spieksma (2009b) reported on the application of integer programming to schedule the Belgian football league for the 2006–2007 and 2007–2008 seasons. Rasmussen (2008) presented a solution approach using a logic-based Benders decomposition and column generation to solve a triple round-robin tournament for the Danish football league. Fiallos et al. (2010) developed an integer programming model solvable by CPLEX, which was used for the first time in 2010 to schedule the double round-robin professional football tournament of Honduras.

The three-phase solution approach described in Sect. 5.2 was based on a first-break, then-schedule decomposition scheme similar to that proposed by Nemhauser and Trick (1998) to solve the problem of scheduling a basketball league. Similar approaches have been used for other sports, although not very many have been implemented and used in practice, like the one reported here.

Bach et al. (2018) described the development of the system that has been in use to schedule the elite series of football in Norway football. They noticed that improved schedules boost attendance and generate positive effects on the local economy. The Norwegian professional football league is a double round robin tournament, where all teams meet each other once at home and once away. To satisfy the stakeholders and thereby create better schedules, a mixed integer programming model is used to schedule the top professional Norwegian football league. The problem was decomposed into multiple stages. The first stage is the assignment of teams to profiles. In the second stage, games are assigned to rounds. This is subject to a set of home and away wishes expressed by the teams, as well as game specific requirements from TV and the league. The approach has been implemented successfully and has been used to develop the schedule in use since the 2018 football season in Norway. The system was particularly useful during the pandemic, when it was also used for rescheduling the games after the interruption of the tournament.

Section 5.3 presents a dynamic approach for scheduling e-sports tournaments using the Swiss system that appeared in Dong et al. (2023).

E-sports were launched by the General Administration of Sport of China as the 99th formal sport in 2003 (Dongsheng et al., 2011). They started as an exhibition sport at the Asian Games (Wade, 2018) in 2018, meaning that the Olympic Committee has recognized them as a formal sports program (Hallmann and Giel, 2018). The eight e-sports medal events scheduled in the 2022 Hangzhou Asian Games (delayed to 2023) have been PUBG Mobile, DOTA 2, Hearthstone, League of Legends, FIFA, Street Fighter V, Arena of Valor, and Dream of the Three Kingdoms 2.

According to a report by Insider Intelligence (Intelligence, 2022), a subscription-based market research company, the monthly viewers for e-sports would reach 29.6 million in 2022, an increase of more than 11.5% compared to 2021. It was also estimated that e-sports enthusiasts would grow by a compound annual rate of 8.1% from 2020 to 2025 and surpass 640 million in 2025 (Newzoo, 2022). The global revenues generated by e-sports in 2021 have reached more than 1 billion US dollars (Statista, 2022), of which the Chinese market accounted for nearly 20%. Moreover, the total prize pools of some popular e-sports tournaments have surpassed those of traditional sports; see (Esports Earnings, 2022; Heath, 2022) for the highest prize pools for e-sports.

Chikish et al. (2019) supported previous evidence emphasizing the complementarity between e-sports and traditional sports. Differences exist between them (Lee and Schoenstedt, 2011). Jenny et al. (2017) noticed that e-sports do not require the same level of physicality associated with standard definitions of sports. Economics and executive ownership aspects of e-sports were discussed by Karhulathi (2017).

Effective rating methods are necessary for dynamic scheduling since we seek to pair teams with similar performance in the competition. Different methods exist for rating the teams in a tournament based on their strengths, such as multi-criteria decision-making (Pradhan and Abdourazakou, 2020) and Colley's ranking method (Colley, 2002; Pasteur, 2010; Stewart, 2013). Führlich et al. (2021) presented alternative pairing rules for Swiss-system chess tournaments based on the computation of maximum weight matchings. Bouzarth et al. (2020) proposed to produce NFL schedules that combine some traditional elements with dynamically scheduled games, optimizing different objectives. Among them are reducing the variability of teams' strength of schedule or minimizing the pairwise comparisons needed to differentiate team quality to make each team's regular season schedule as fair as possible. These concerns are similar to those considered in the dynamic scheduling approach for e-sports tournaments described in Sect. 5.3, where two computational experiments were reported. Real-life data of the 2020 Honor of Kings World Champion Cup (KCC2020) were collected from Liquipedia (2021). The accuracy and the suitability of the Swiss system can be especially appealing for emerging sports such as e-sports, where the tournament designs are not yet solidified, corroborating the views of other authors such as Sziklai et al. (2022).

The dynamic scheduling approach for e-sports tournaments is compatible with other ranking methods and sports. If traveling over long distances is not a key aspect of the tournament (or if there is enough time to organize the logistics between consecutive rounds), there is no need to determine the games in each round before the start of the tournament. The organizers can use dynamic scheduling to improve the competition's suspense and fairness. Thus the proposed approach for e-sports can be extended to traditional sports with similar properties. This situation is the case of the European Champions League, which has decided to use the Swiss system for the 2024-25 season. The number of teams will rise from 32 to 36. Teams are guaranteed a minimum of eight games with the new design. As a consequence, the tournament almost doubles in size, from a total of 125 games to 225. The top teams should be able to play more meaningful games against their main rivals (ESPN, 2021).

Table 5.9 presents a collection of applications of sports scheduling. For each of them, we indicate the sports discipline involved, the type of problem that motivated the application, the solution approach adopted by the authors, the country of the competition, and the reference where it is reported. We notice that all these applications are reported in the literature as real-life cases. However, despite considering real problems, only some of them were used in practice to schedule specific seasons of the tournaments considered.

Table 5.9 Applications of sports scheduling (MIP: mixed integer programming; MINLP: mixed integer nonlinear programming; CP: constraint programming; LS: local search, TS: tabu search; SA: simulated annealing; GA: genetic algorithm; MA: memetic algorithm; SAT: satisfiability)

Sport	Problem	Country	References	Approach
Baseball	Tournament scheduling, relaxed	USA	Cain (1977)	Heuristic
	Tournament scheduling, compact	USA	Russell and Leung (1994)	Heuristic
	Tournament scheduling, compact	Japan	Hoshino and Kawarabayashi (2011a)	Graph methods
	Tournament scheduling, compact	Japan and USA	Hoshino and Kawarabayashi (2011b)	Graph methods
	Umpire scheduling	USA	Trick et al. (2012)	SA, MIP, and CP
Basketball	Tournament scheduling, compact	Australia	de Werra et al. (1990)	Graph methods
	Tournament scheduling, relaxed	USA	Campbell and Chen (1976)	Heuristic
	Tournament scheduling, relaxed	USA	Ball and Webster (1977)	MIP and heuristics
	Tournament scheduling, relaxed	USA	Bean and Birge (1980)	Heuristic
	Tournament scheduling, compact	USA	Nemhauser and Trick (1998)	MIP
	Tournament scheduling, compact	USA	Henz (2001)	CP
	Tournament scheduling, compact	Czech Republic	Fronček (2001)	Graph methods
	Tournament scheduling, compact	USA	Zhang (2002)	SAT solver
	Tournament scheduling, compact	USA	van Voorhis (2002)	MIP
	Tournament scheduling, compact	USA	van Voorhis (2005)	MIP
	Tournament scheduling, relaxed	New Zealand	Wright (2006)	SA
	Tournament scheduling, compact	Germany	Westphal (2014)	MIP
Cricket	Umpire scheduling	England	Wright (1991)	LS
	Tournament scheduling, relaxed	World Cup	Armstrong and Willis (1993)	Spreadsheet
	Tournament scheduling, relaxed	England	Wright (1992)	SA
	Tournament scheduling, relaxed	England	Wright (1994)	TS

Table 5.9 (continued)

Sport	Problem	Country	References	Approach
	Tournament scheduling, relaxed	Australia	Willis and Terrill (1994)	SA
	Tournament scheduling, relaxed	New Zealand	Wright (2005)	SA
	Tournament scheduling, relaxed	England	Wright (2018)	SA
Football (American)	Tournament scheduling, compact	USA	Dinitz and Fronček (2000)	Graph methods
Football (Australian)	Tournament scheduling, compact	Australia	Kyngäs et al. (2017)	Heuristic
Football (Canadian)	Tournament scheduling, compact	Canada	Kostuk and Willoughby (2012)	MIP
Football (Indoor)	Tournament scheduling, relaxed	Belgium	Van Bulck et al. (2019)	TS
Football (Soccer)	Tournament scheduling, compact	Netherlands	Schreuder (1992)	MIP
	Tournament scheduling, compact	Austria and Germany	Bartsch et al. (2006)	MIP and heuristic
	Tournament scheduling, compact	Italy	Della Croce and Oliveri (2006)	MIP
	Tournament scheduling, compact	Chile	Noronha et al. (2007)	MIP
	Referee assignment	USA	Duarte et al. (2007b)	ILS
	Referee assignment	USA	Duarte et al. (2007a)	MIP and ILS
	Tournament scheduling, compact	Denmark	Rasmussen (2008)	MIP and CP
	Referee assignment	USA	Duarte and Ribeiro (2008)	Bi-objective LS
	Tournament scheduling, compact	Chile	Durán et al. (2007),	MIP and CP
			Durán et al. (2009)	MIP
	Tournament scheduling, compact	Belgium	Goossens and Spieksma (2009b)	MIP
	Tournament scheduling, compact	Norway	Flatberg et al. (2009)	MIP
	Tournament scheduling, compact	Honduras	Fiallos et al. (2010)	MIP
	Tournament scheduling, compact	Brazil	Ribeiro and Urrutia (2007b), Ribeiro and Urrutia (2010), Ribeiro and Urrutia (2012)	MIP

(continued)

Table 5.9 (continued)

	Tournament scheduling, compact	Chile	Durán et al. (2012)	MIP
	Tournament scheduling, compact	Ecuador	Recalde et al. (2013)	MIP
	Referee assignment	Chile	Alarcón et al. (2014)	MIP
	Tournament scheduling, compact	World Cup qualifiers	Durán et al. (2017)	MIP
	Tournament scheduling, compact	Norway	Bach et al. (2018)	MIP and GRASP
	Tournament scheduling, compact	Turkey	Günneç and Demir (2018)	Heuristics
	Tournament scheduling, compact	Belgium	Goossens (2018)	MIP
	Tournament scheduling, compact	Ecuador	Durán et al. (2022)	MIP
	Match-day assignment	Turkey	Göçgün and Bakır (2022)	MINLP
Handball	Tournament scheduling, compact	Sweden	Larson and Johansson (2014)	MIP and CP
Ice hockey	Tournament design and scheduling, relaxed	Canada and USA	Fleurent and Ferland (1993)	MIP
	Tournament scheduling, relaxed	Canada and USA	Costa (1995)	Evolutionary TS
	Tournament scheduling, relaxed	Finland	Kyngäs and Nurmi (2009), Nurmi et al. (2015)	Hybrid heuristics
	Tournament design and scheduling, relaxed	Finland	Nurmi et al. (2014)	Population-based LS
Rugby	Tournament scheduling, compact	Wales	Lewis and Thompson (2011)	LS
	Tournament scheduling, compact	New Zealand	Johnston and Wright (2014)	SA
Softball	Tournament scheduling, compact	USA	Saur et al. (2012)	MIP
Table tennis	Tournament scheduling, relaxed	Germany	Schönberger et al. (2004)	MA and CP
	Tournament scheduling, relaxed	Germany	Knust (2010)	MIP, GA and LS
Tennis	Tournament scheduling, compact	Italy	Della Croce et al. (1999)	Heuristics
	Umpire scheduling	USA	Farmer et al. (2007)	MIP and SA
Volleyball	Tournament scheduling, compact	Argentina	Bonomo et al. (2012)	TS
	Tournament scheduling	Italy	Cocchi et al. (2018)	MIP

References

Alarcón F, Durán G, Guajardo M (2014) Referee assignment in the Chilean football league using integer programming and patterns. Int Trans Oper Res 21:415–438

Anderson I (1999) Balancing carry-over effects in tournaments. In: Holroyd F, Quinn K, Rowley C, Webb B (eds) Combinatorial designs and their applications. Chapman & Hall, Boca Raton, pp 1–16

Armstrong J, Willis RJ (1993) Scheduling the cricket World Cup – a case-study. J Oper Res Soc 44:1067–1072

Bach L, Kloster O, Ribeiro CC (2018) Scheduling professional football leagues. In: Conference handbook of the 29th European conference on operational research, p 398

Ball BC, Webster DB (1977) Optimal scheduling for even-numbered team athletic conferences. AIIE Trans 9:161–169

Bartsch T, Drexl A, Kröger S (2006) Scheduling the professional soccer leagues of Austria and Germany. Comput Oper Res 33:1907–1937

Bean JC, Birge JR (1980) Reducing travelling costs and player fatigue in the National Basketball Association. Interfaces 10:98–102

Bonomo F, Cardemil A, Durán G, Marenco J, Sabán D (2012) An application of the traveling tournament problem: the argentine volleyball league. Interfaces 42:245–259

Bouzarth EL, Cromer AW, Fravel WJ, Grannan BC, Hutson KR (2020) Dynamically scheduling NFL games to reduce strength of schedule variability. J Sports Anal 6:281–293

Cain WO (1977) The computer-aided heuristic approach used to schedule the Major league baseball clubs. In: Ladany SP, Machol RE (eds) Optimal strategies in sports. North Holland, Amsterdam, pp 33–41

Campbell RT, Chen DS (1976) A minimum distance basketball scheduling problem. In: Machol RE, Ladany SP, Morrison DG (eds) Management science in sports. Studies in the management sciences, vol 4. North-Holland, Amsterdam, pp 15–25

Chikish Y, Carreras M, García J (2019) esports: A new era for the sports industry and a new impulse for the research in sports (and) economics. In: Sports (and) economics. Spanish Savings Banks Foundation, Madrid, pp 477–508

Cocchi G, Galligari A, Nicolino F, Schoen F, Sciandrone M, Piccialli V (2018) Scheduling the Italian national volleyball tournament. Interfaces 48:271–284

Colley WN (2002) Colley's bias free college football ranking method: The Colley matrix explained. Technical Report, Princeton University. Online reference at https://www.colleyrankings.com/matrate.pdf, last access on July 30, 2022

Costa D (1995) An evolutionary tabu search algorithm and the NHL scheduling problem. INFOR 33:161–178

Costa FN, Urrutia S, Ribeiro CC (2008) An ILS heuristic for the traveling tournament problem with fixed venues. In: Burke EK, Gendreau M (eds) Proceedings of the 7th international conference on the practice and theory of automated timetabling

Costa FN, Urrutia S, Ribeiro CC (2012) An ILS heuristic for the traveling tournament problem with predefined venues. Ann Oper Res 194:137–150

Della Croce F, Oliveri D (2006) Scheduling the Italian football league: an ILP-based approach. Comput Oper Res 33:1963–1974

Della Croce F, Tadei R, Asioli P (1999) Scheduling a round robin tennis tournament under courts and players availability constraints. Ann Oper Res 92:349–361

Dinitz JH (1996) Starters. In: Colbourn CJ, Dinitz JH (eds) The CRC handbook of combinatorial designs. The CRC Press series on discrete mathematics and its applications. CRC Press, Boca Raton, pp 467–473

Dinitz JH, Fronček RE (2000) Scheduling the XFL. Congressus Numerantium 147:5–15

Dong ZL, Ribeiro CC, Xu F, Zamora A, Ma Y, Jing K (2023) Dynamic scheduling of e-sports tournaments. Trans Res Part E 169:102,988. https://doi.org/10.1016/j.tre.2022.102988

Dongsheng Y, Xiaohang Y, Daofeng K (2011) The present situation and development trend of e-sports games in China. In: 2011 international conference on future computer science and education. IEEE, pp 384–386

Duarte AR, Ribeiro CC (2008) Referee assignment in sports leagues: approximate and exact multi-objective aproaches. In: 19th international conference on multiple criteria decision making, pp 58–60

Duarte AR, Ribeiro CC, Urrutia S (2007a) A hybrid ILS heuristic to the referee assignment problem with an embedded MIP strategy. In: Bartz-Beielstein T, Aguilera MJB, Blum C, Naujoks B, Roli A, Rudolph G, Sampels M (eds) Hybrid metaheuristics. Lecture notes in computer science, vol 4771. Springer, Berlin, pp 82–95

Duarte AR, Ribeiro CC, Urrutia S, Haeusler EH (2007b) Referee assignment in sports leagues. In: Burke EK, Rudová H (eds) Practice and theory of automated timetabling VI. Lecture notes in computer science, vol 3867. Springer, Berlin, pp 158–173

Durán G, Guajardo M, Miranda J, Sauré D, Souyris S, Weintraub A, Wolf R (2007) Scheduling the Chilean soccer league by integer programming. Interfaces 37:539–552

Durán G, Guajardo M, Weintraub A, Wolf R (2009) O.R. & Soccer: scheduling the Chilean league using mathematical programming. OR/MS Today 36:42–47

Durán G, Guajardo M, Wolf-Yadlin R (2012) Operations research techniques for scheduling Chile's second division soccer league. Interfaces 42:273–285

Durán G, Guajardo M, Sauré D (2017) Scheduling the South American qualifiers to the 2018 FIFA World Cup by integer programming. Eur J Oper Res 262:1109–1115

Durán G, Guajardo M, Zamorano G (2022) Mathematical models for rescheduling Ecuador's 2020 professional football league season disrupted by COVID-19. Soc Econ 44:360–377

ESPN (2021) Champions League Swiss model revamp: All you need to know. Online reference at https://www.espn.com/soccer/uefa-champions-league/story/4366699/champions-league-swiss-model-revamp-all-you-need-to-know, last access on December 2, 2022

Esports Earnings (2022) E$PORTS EARNINGS. Online reference at https://www.esportsearnings.com/tournaments, last access on August 7, 2022

Farmer A, Smith JS, Miller LT (2007) Scheduling umpire crews for professional tennis tournaments. Interfaces 37:187–196

Führlich P, Cseh A, Lenzner P (2021) Improving ranking quality and fairness in Swiss-system chess tournaments. Online reference at arXiv:2112.10522v1, last visited on January 30, 2023

Fiallos J, Pérez J, Sabillón F, Licona M (2010) Scheduling soccer league of Honduras using integer programming. In: Johnson A, Miller J (eds) Proceedings of the 2010 industrial engineering research conference, San Carlos

Flatberg T, Nilssen E, Stolevik M (2009) Scheduling the topmost fotball leagues of Norway. In: Book of abstracts of the 23rd European conference on operational research, p 240

Fleurent C, Ferland JA (1993) Allocating games for the NHL using integer programming. Oper Res 41:649–654

Frončck D (2001) Scheduling the Czech national basketball league. Congressus Numerantium 153:5–24

Göçgün Y, Bakır NO (2022) Optimal matchday schedule for Turkish professional soccer league using nonlinear binary integer programming. Int J Optim Control Theor Appl 12:113–127

Glover F (1996) Ejection chains, reference structures and alternating path methods for traveling salesman problems. Discrete Appl Math 65:223–253

Glover F, Laguna M (1997) Tabu search. Kluwer, Alphen aan den Rijn

Glover F, Laguna M, Martí R (2000) Fundamentals of scatter search and path relinking. Control Cybern 29(3):653–684

Günneç D, Demir E (2018) Fair fixture: minimizing carry-over effects in football leagues. J Indust Manag Optim 15:1565–1577

Goodbread C (2010) SEC aims to fix schedule problem by month's end. Online reference at https://eu.tuscaloosanews.com/story/news/2010/04/09/sec-aims-to-fix-schedule-problem-by-months-end/27948131007/, last visited on June 6, 2022

Goossens DR (2018) Optimization in sports league scheduling: experiences from the Belgian Pro League soccer. Commun Comput Inf Sci 884:3–19
Goossens DR, Spieksma FCR (2009a) Does the carry-over effect exist? In: Book of abstracts of the 23rd European conference on operational research. Bonn, p 288
Goossens DR, Spieksma FCR (2009b) Scheduling the Belgian soccer league. Interfaces 39:109–118
Guedes ACB, Ribeiro CC (2011) A heuristic for minimizing weighted carry-over effects in round robin tournaments. J Sched 14:655–667
Gutin G, Punnen P (eds) (2002) The traveling salesman problem and its variations. Kluwer, Alphen aan den Rijn
Hallmann K, Giel T (2018) esports – competitive sports or recreational activity? Sports Manag Rev 21:14–20
Hansen P, Mladenović N (2002) Developments of variable neighborhood search. In: Ribeiro CC, Hansen P (eds) Essays and surveys in metaheuristics. Kluwer, Alphen aan den Rijn, pp 415–439
Heath J (2022) The top 10 highest prize pools in esports. Online reference at https://dotesports.com/general/news/biggest-prize-pools-esports-14605, last access on August 7, 2022
Henz M (2001) Scheduling a major college basketball conference - revisited. Oper Res 49:163–168
Henz M, Müller T, Thiel S (2004) Global constraints for round robin tournament scheduling. Eur J Oper Res 153:92–101
Hoshino R, Kawarabayashi K (2011a) A multi-round generalization of the traveling tournament problem and its application to Japanese baseball. Eur J Oper Res 215:481–497
Hoshino R, Kawarabayashi K (2011b) Scheduling bipartite tournaments to minimize total travel distance. J Artif Intell Res 42:91–124
Insiderintelligence (2022) Esports ecosystem in 2022: key industry companies, viewership growth trends, and market revenue stats. Online reference at https://www.insiderintelligence.com/insights/esports-ecosystem-market-report/, last access on August 7, 2022.
Januario T, Urrutia S (2016) A new neighborhood structure for round robin scheduling problems. Comput Oper Res 70:127–139
Januario T, Urrutia S, Ribeiro CC, de Werra D (2016) Edge coloring: a natural model for sports scheduling. Eur J Oper Res 254:1–8
Jenny SE, Manning RD, Keiper MC, Olrich TW (2017) Virtual(ly) athletes: where esports fit within the definition of "sport". Quest 69:1–18
Johnston M, Wright M (2014) Prior analysis and scheduling of the 2011 Rugby Union ITM Cup in New Zealand. J Oper Res Soc 65:1292–1300
Karhulathi V (2017) Reconsidering Esport: economics and executive ownership. Phys Cult Sport Stud Res 74:43–53
Kirkman TP (1847) On a problem in combinations. Camb Dublin Math J 2:191–204
Knust S (2010) Scheduling non-professional table-tennis leagues. Eur J Oper Res 200:358–367
Kostuk KJ, Willoughby KA (2012) A decision support system for scheduling the Canadian football league. Interfaces 42:286–295
Kyngäs J, Nurmi K (2009) Scheduling the Finnish major ice hockey league. In: Proceedings of the 2009 IEEE symposium on computational intelligence in scheduling. IEEE, pp 84–89
Kyngäs J, Nurmi K, Kyngäs N, Lilley G, Salter T, Goossens DR (2017) Scheduling the Australian football league. J Oper Res Soc 68:973–982
Lambrechts E, Ficker AMC, Goossens DR, Spieksma FCR (2018) Round-robin tournaments generated by the circle method have maximum carry-over. Math Program 172:277–302
Larson J, Johansson M (2014) Constructing schedules for sports leagues with divisional and round-robin tournaments. J Quant Anal Sports 10:119–129
Lawler EL, Lenstra JK, Rinnooy Kan AHG, Shmoys DB (eds) (1985) The traveling salesman problem: a guided tour of combinatorial optimization. Wiley, New York
Lee D, Schoenstedt LJ (2011) Comparison of esports and traditional sports consumption motives. ICHPER-SD J Res 6:39–44
Lewis R, Thompson J (2011) On the application of graph colouring techniques in round-robin sports scheduling. Comput Oper Res 38:190–204

Liquipedia (2021) Honor of Kings World Champion Cup/2020/Group Stage. Online reference at https://liquipedia.net/arenaofvalor/Honor_of_Kings_World_Champion_Cup/2020/Group_Stage, last access on August 5, 2022
Miyashiro R, Matsui T (2006) Minimizing the carry-over effects value in a round robin tournament. In: Burke EK, Rudová H (eds) Proceedings of the 6th international conference on the practice and theory of automated timetabling, pp 460–473
Mladenovic N, Hansen P (1997) Variable neighborhood search. Comput Oper Res 34:1097–1100
Nemhauser GL, Trick MA (1998) Scheduling a major college basketball conference. Oper Res 46:1–8
Newzoo (2022) Newzoo's global esports & live streaming market report 2022. Online reference at https://newzoo.com/insights/trend-reports/newzoo-global-esports-live-streaming-market-report-2022-free-version, last access on August 14, 2022.
Noronha TF, Ribeiro CC, Durán G, Souyris S, Weintraub A (2007) A branch-and-cut algorithm for scheduling the highly-constrained Chilean soccer tournament. In: Burke EK, Rudová H (eds) Practice and theory of automated timetabling VI. Lecture notes in computer science, vol 3867. Springer, Berlin, pp 174–186
Nurmi K, Goossens DR, Bartsch T, Bonomo F, Briskorn D, Duran G, Kyngäs J, Marenco J, Ribeiro CC, Spieksma FCR, Urrutia S, Wolf-Yadlin R (2010) A framework for scheduling professional sports leagues. IAENG Trans Eng Technol 5:14–28
Nurmi K, Goossens DR, Kyngäs J (2014) Scheduling a triple round robin tournament with minitournaments for the Finnish national youth ice hockey league. J Oper Res Soc 65:1770–1779
Nurmi K, Kyngäs J, Goossens DR, Kyngäs N (2015) Scheduling the finnish major ice hockey league using the peast algorithm. In: Yang GC, Ao SI, Huang X, Castillo O (eds) Transactions on engineering technologies: international multiconference of engineers and computer scientists 2014. Springer, Berlin, pp 155–168
Pasteur RD (2010) Extending the Colley method to generate predictive football rankings. In: Gallian JA (ed) Mathematics and sports, American mathematical society, chap 10, pp 117–130
Pradhan S, Abdourazakou Y (2020) "Power ranking" professional circuit esports teams using multi-criteria decision-making (MCDM). J Sports Anal 6(1):61–73
Rasmussen RV (2008) Scheduling a triple round robin tournament for the best Danish soccer league. Eur J Oper Res 185:795–810
Recalde D, Torres R, Vaca P (2013) Scheduling the professional Ecuadorian football league by integer programming. Comput Oper Res 40:2478–2484
Ribeiro CC, Urrutia S (2007a) Heuristics for the mirrored traveling tournament problem. Eur J Oper Res 179:775–787
Ribeiro CC, Urrutia S (2007b) Scheduling the Brazilian soccer tournament with fairness and broadcast objectives. In: Burke EK, Rudová H (eds) Practice and theory of automated timetabling VI. Lecture notes in computer science, vol 3867. Springer, Berlin, pp 147–157
Ribeiro CC, Urrutia S (2010) Soccer scheduling goaaaaal! OR/MS Today 37(2):54–59
Ribeiro CC, Urrutia S (2012) Scheduling the Brazilian soccer tournament: solution approach and practice. Interfaces 42:260–272
Russell KG (1980) Balancing carry-over effects in round robin tournaments. Biometrika 67:127–131
Russell RA, Leung JM (1994) Devising a cost effective schedule for a baseball league. Oper Res 42:614–625
Saur MC, Starr K, Husted M, Newman AM (2012) Scheduling softball series in the Rocky Mountain Athletic Conference. Interfaces 42:296–309
Schönberger J, Mattfeld DC, Kopfer H (2004) Memetic algorithm timetabling for non-commercial sport leagues. Eur J Oper Res 153:102–116
Schreuder JAM (1992) Combinatorial aspects of construction of competition Dutch professional football leagues. Discrete Appl Math 35:301–312

Statista (2022) eSports market revenue worldwide from 2020 to 2025. Online reference at https://www.statista.com/statistics/490522/global-esports-market-revenue/, last access on October 25, 2022.

Stewart MJ (2013) A proposed general rating system for which the Colley matrix rating system is a special case. J Quant Anal Sports 9:179–185

Sziklai BR, Biró P, Csató L (2022) The efficacy of tournament designs. Comput Oper Res 144:105,821

Trick MA (2000) A schedule-then-break approach to sports timetabling. In: Burke EK, Erben W (eds) Selected papers from the third international conference on practice and theory of automated timetabling III. Lecture notes in computer sciences, vol 2079. Springer, Berlin, pp 242–253

Trick MA, Yildiz H, Yunes T (2012) Scheduling Major League Baseball umpires and the traveling umpire problem. Interfaces 42:232–244

Van Bulck D, Goossens DR, Spieksma F (2019) Scheduling a non-professional indoor football league: a tabu search based approach. Ann Oper Res 275:715–730

van Voorhis TV (2002) Highly constrained college basketball scheduling. J Oper Res Soc 53:603–609

van Voorhis TV (2005) College basketball scheduling with travel swings. Comput Industrial Eng 48:163–172

Wade S (2018) Esports are debuting at the Asian Games. Could the Olympics be next? Online reference at https://www.denverpost.com/2018/08/27/esports-asian-games-olympics/, last access on August 3, 2022

de Werra D (1980) Geography, games and graphs. Discrete Appl Math 2:327–337

de Werra D (1981) Scheduling in sports. In: Hansen P (ed) Studies on graphs and discrete programming. Annals of discrete mathematics, vol 11. North-Holland, Amsterdam, pp 381–395

de Werra D, Jacot-Descombes L, Masson P (1990) A constrained sports scheduling problem. Discrete Appl Math 26:41–49

Westphal S (2014) Scheduling the German basketball league. Interfaces 44:498–508

Willis RJ, Terrill BJ (1994) Scheduling the Australian state cricket season using simulated annealing. J Oper Res Soc 45:276–280

Wright M (1992) A fair allocation of county cricket opponents. J Oper Res Soc 43:195–201

Wright M (1994) Timetabling county cricket fixtures using a form of tabu search. J Oper Res Soc 45:758–770

Wright M (2005) Scheduling fixtures for New Zealand cricket. IMA J Manag Math 16:99–112

Wright M (2006) Scheduling fixtures for basketball New Zealand. Comput Oper Res 33:1875–1893

Wright M (2018) Scheduling an amateur cricket league over a nine-year period. J Oper Res Soc 69:1854–1862

Wright MB (1991) Scheduling English cricket umpires. J Oper Res Soc 42:447–452

Zhang H (2002) Generating college conference basketball schedules by a SAT solver. In: Proceedings of the fifth international symposium on the theory and applications of satisfiability testing, pp 281–291

Epilogue

After reaching the end of this book and revisiting the content of all its chapters, all our goals have been attained: we are glad to realize that sports scheduling has been a fantastic illustration of the various uses and applications of Operations Research.

Considering and handling many problems characterized by specific objectives, constraints, and requirements has been a typical modeling exercise, the starting point of every Operations Research study. It is delightful to see that a comprehensive sample of fundamental Operations Research techniques and models has appeared over the preceding chapters. They range from elementary graph concepts to intricate integer programming models and an impressive collection of heuristic procedures to get real solutions that have been used in concrete situations of scheduling sports tournaments.

It is also highly stimulating to realize that the existence of efficient procedures to solve the current scheduling problems will undoubtedly encourage many sports leagues to raise new challenging problems. These problems and applications will, in turn, motivate the development of new models and new solution techniques.

So, the game is not over!

C. C. Ribeiro et al., *Combinatorial Models for Scheduling Sports Tournaments*,
EURO Advanced Tutorials on Operational Research,
https://doi.org/10.1007/978-3-031-37283-4

MIX
Papier aus verantwortungsvollen Quellen
Paper from responsible sources
FSC® C105338

If you have any concerns about our products,
you can contact us on
ProductSafety@springernature.com

In case Publisher is established outside the EU,
the EU authorized representative is:
Springer Nature Customer Service Center GmbH
Europaplatz 3, 69115 Heidelberg, Germany

Printed by Libri Plureos GmbH
in Hamburg, Germany